Charles W. Holbrook

Lunar Tellurian

Charles W. Holbrook

Lunar Tellurian

ISBN/EAN: 9783337371531

Printed in Europe, USA, Canada, Australia, Japan

Cover: Foto ©berggeist007 / pixelio.de

More available books at **www.hansebooks.com**

TEACHER'S MANUAL

EXPLAINING THE USE OF

CHAS. W. HOLBROOK'S

LUNAR TELLURIAN.

—✳—

PRINTED FOR THE AUTHOR BY
THE CASE, LOCKWOOD & BRAINARD COMPANY, HARTFORD, CONN.

PREFACE.

TEACHING is to some extent, as the genial Dickens makes Mantellini say, "one demd horrid grind."

The wear and tear, the dull routine, the hopelessness of dull minds, the mechanical aspect of the teacher's life, the hard work and poor pay of the noblest calling; these facts have been constantly uppermost in my mind, lending zest to an endeavor to *help* the teacher. I should consider the mere inventing, patenting, making, and selling of my Tellurian an ignoble aim unless the desire of an income were in some measure transfused by an attempt to lighten the burdens of teaching. Surely to inspire a mind with zeal for self-improvement is a lofty act. To effect a permanent lodgment in a human brain an *idea* must be illustrated by a mechanical fact when possible. To govern and administer a school is a rare talent — developed in many teachers, latent, however, in the most cases. The idea of law, system, *obedience*, has its origin in nature — there and there only can we find the evolution of the idea physically illustrated in what we call the phenomena of nature. When a teacher can close his eyes, transplant himself into space, and *see* the earth pursuing the mysterious round of rotation and orbit — the tiny moon swimming and circling in her endless journey of subjective obedience to the greater orb; all facts of primary astronomy become instantly apparent. He has then but one use for this mechanical instrument, viz.: to *prove* to the undeveloped imagination the invisible facts as seen by himself through the visual process of mental insight. Enthusiasm supplants the mere sense of routine duty, original illustration by means of the pencil, chalk, diagrams, etc., occur as the pupil catches the infection and asks questions, and it goes without saying that one hour of genuine enthusiasm in a class is worth a whole term of dragging, monotonous repetition of words.

For *their own sake* and the benefit of the class, then, the writer asks teachers to give this little book some real attention. It makes no pretensions to literary or any other merit than as a help to the tired teacher, whose aims and aspirations and disappointments and successes have his full sympathy.

CHARLES W. HOLBROOK.

(3)

INDEX.

(4)

A NEW GLOBE.

TELLURIAN AND LUNARIAN IN ONE.

Invented and made by Charles W. Holbrook.

PATENTED MARCH 6, 1888.

MANUAL OF EXPLANATION WITH EACH INSTRUMENT.

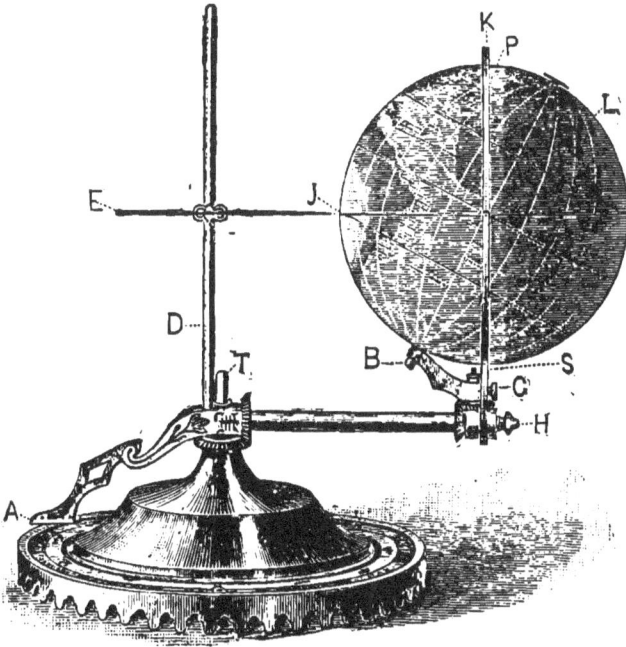

All phenomena of Sun, Earth, and Moon elucidated by this simple, accurate globe.

Not a complicated, delicate machine to get out of order and be set aside for dust and flies.

(5)

All parts visible and interchangeable should any become disabled by accident.

Not a hidden bolt, bar, or gear; no secreted belts to break or loosen.

Easily adjusted, ready for use, cannot be set up wrongly for study.

Not a long course of practice required to understand it.

Any one can use it and with the aid of the Manual, understand natural causes and effects at sight.

Not an instrument that will catch, hitch, break down and stop — embarrassing a teacher before his class.

Can be absolutely depended upon. The only globe which illustrates, with any degree of accuracy, the courses of the moon as verified by almanacs; showing phases, eclipses, nodes; when, how, and why they occur. Has a transparent shade for the dark side of the earth, also a shade for the moon, operating exactly as in nature. Affords interesting study for class room and library.

This new Tellurian, Lunarian, Globe, three instruments in one, illustrates, with precision, over one hundred astronomical conditions; elucidating phenomenal or *apparent* facts and deducing *real* or noumenal facts. I ask your attention to the manual written, especially for this globe, in the ordinary language of conversation; avoiding, as far as possible, the technical terms of astronomy. The greatest obstacle to the general use of such a globe is the want of familiarity, on the part of teachers and parents, with the *mechanics* of our solar system. They should understand the simple arrangement by which our world maintains its relations to the sun, and the moon to the earth.

The *mathematics* of astronomy is a higher and more abstract branch of study known only to the few.

The *mechanical* facts are easily understood and instantly; without any abstruse calculations or great effort of the imagination. Among the many the following list is given of the mechanical conditions existing at the different times and places of the earth in its annual six hundred million miles journey around the sun.

A.— Daily Motion of the Earth on its Axis, illustrating the following facts, viz.:

1. That the sun, moon, and stars do not have the movements apparent to the eyes, but that the earth has a movement exactly opposite to what appears in the sky.

2. The division of the earth's surface into sections of meridian and parallel and why the astronomers were obliged to do this.

3. Cause of day and night and twilight.

4. Passage of the hours of a clock dial entirely around the earth every day. In plainer words, the fact that NOON is a condition of a constant and endless duration; it is simply the moment when your locality is brought nearest the sun by the earth's daily rotation.

5. The time of day by your clock is earlier than at places east of you at the rate of one hour to each fifteen degrees; and later than at places westward.

6. Quantity and conditions of heat emitted by the sun — effects of heat upon the earth — difference between vertical and oblique solar rays and consequent division of the map into Zones.

7. A device for mounting the globe so as to show the earth with a perpendicular axis, like Jupiter, and explanations why seasons would never change, but

climate would vary according to distance from equator; days and nights would be 12 hours each, always. This experiment is vital to a clear understanding of the real significance of an axis inclined as is that of our earth.

B.—Inclination of the Earth's Axis.
The plane of the Ecliptic, why so called and its significance. The earth's Orbit or Annual Path. Two primal facts bearing vast results illustrated and explained.

8. Constellations and signs as given in astronomy. Their uses and value as sign-posts.

9. Why the sun is high in summer and low in winter, though it never moves.

10. Sun "crossing the line."

11. Why the day and night are each 12 hours on March 23d and September 23d, all over the world and at no other times.

12. Why the sun rises at 6 and sets at 6, every day of the year, at the equator.

13. Why days are not of the same length at different places for the same time.

14. Why days are not uniform at the same places for different times.

15. Full and exhaustive study on variations, durations, and causes of time, with interesting experiments.

16. The midnight sun at the Arctic Circle. A polar day and night, each six months long, with a description of the sun's apparent course seen by a man standing on the north pole.

17. Difference between a sidereal and solar day.

18. Description of the sun's constitution, size, and importance.

19. Physical significance of atmosphere — how its loss would depopulate the earth in a moment.

20. The theories of Milton and others about the inclination of the axis.

21. The axis of the earth always true to the same direction, called "parallelism."

22. Zenith and Nadir, when and where the sun is directly above your head, its distance from that Zenith for any day in the year.

23. Change of seasons — why we do not have a uniform summer as they do at the equator.

24. Description of the four seasons, showing why they change at distances from the equator without varying there.

25. Why it is summer at all places south of the equator during our northern winter, and vice-versa.

C. — Climate.

A chapter explaining with the aid of the globe.

26. Why the climates of different places on the same parallel at the same time are so unlike.

27. Humboldt's researches.

28. Alternations of seasons.

29. Local variations.

30. Explanation of the red and blue lines on the globe.

31. Effect of mountains, plains, distance from sea ; height of land levels, etc., etc.

32. Polar and equatorial currents of water and air constituting the ocean currents and trade winds.

33. The Gulf Stream and its effect upon climate.

34. The myrtle blooms in winter in Ireland, and why.

35. Sun's distance from the earth at winter and summer.

36. Greater accumulation of ice at the Antarctic than at the Arctic region.

37. How this causes predominating currents of water and air thence to the equator.

38. Civilization the result of climate and slow northward march of human achievement, since the rise of Thebes, identical to the northward advance of the temperate zone.

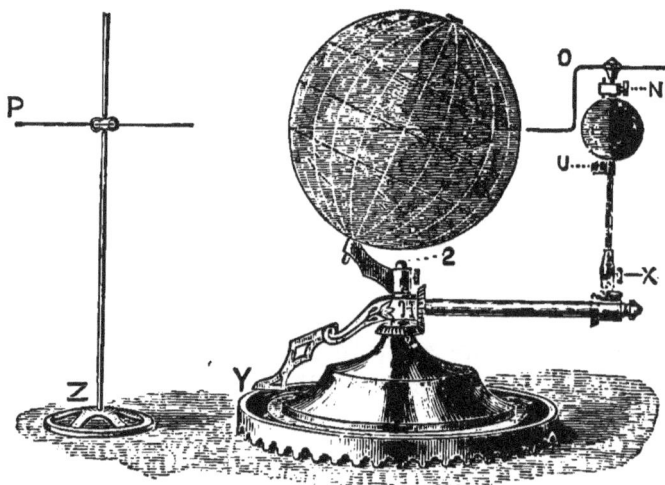

The Lunarian illustrates all facts relating to the Moon's courses, some of which may be given, viz. :

39. The moon's actual motion on its axis.

40. Face of the moon always toward the earth.

41. One-half always bright, whether visible to the earth or not — the other half dark by rotation into its own shadow.

42. The moon's day and night each 15 times as long as our own.

43. Difference between the moon's sidereal and solar days.

44. The inclination of the moon's orbit, or monthly path, to the plane of the earth's orbit or yearly path.

Illustrated by no other instrument with any degree of exactness.

45. Phases of the moon — causes of a "full," "quarter," "crescent," and " new " moon.

46. A bright moon with a dark shade, showing the actual graduation of phases according to nature.

47. Why the moon is "high" in winter and "low" in summer.

48. Nodes of the moon — mystery made clear.

49. Conjunction, quadrature, and opposition.

50. Why the moon must be at or near a node to cause an eclipse.

51. Why solar and lunar eclipses do not occur every month, during the moon's journey around the earth.

52. Total, Annular, Partial eclipses of the sun — their simple causes and effects.

53. Moon's distance from the earth, called Perigee and Apogee, their various effects upon eclipses.

54. A journey of the moon as described in the almanacs, traced, illustrated, and explained, including actual eclipses as they occurred.

55. Tides, their causes and effects.

Note to Purchasers.

If you receive the globe with the *moon screwed on the arm*, all you need to do to complete the adjustment of the lunarian is to set the globe upon the central part,

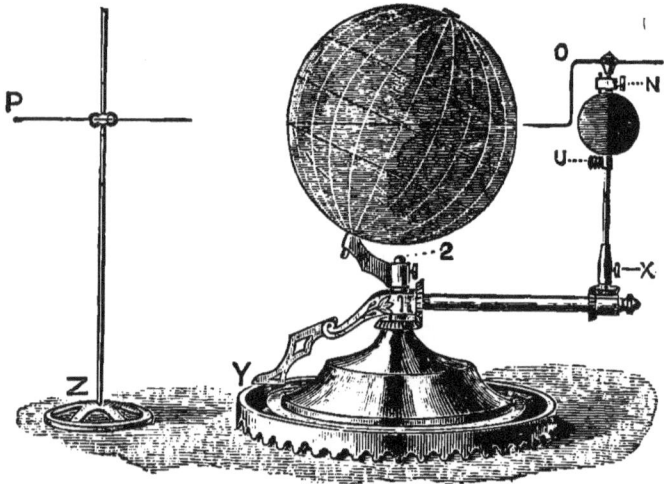

arranging the other parts as seen in the cut. *To take off the moon*, loosen thumb-screw X, turn to the left by grasping at N. When you adjust the moon, follow directions given under the head of " Lunar Tellurian," in the manual, taking care to have X loose and N tightly screwed up.

CHAS. W. HOLBROOK'S
LUNAR TELLURIAN.

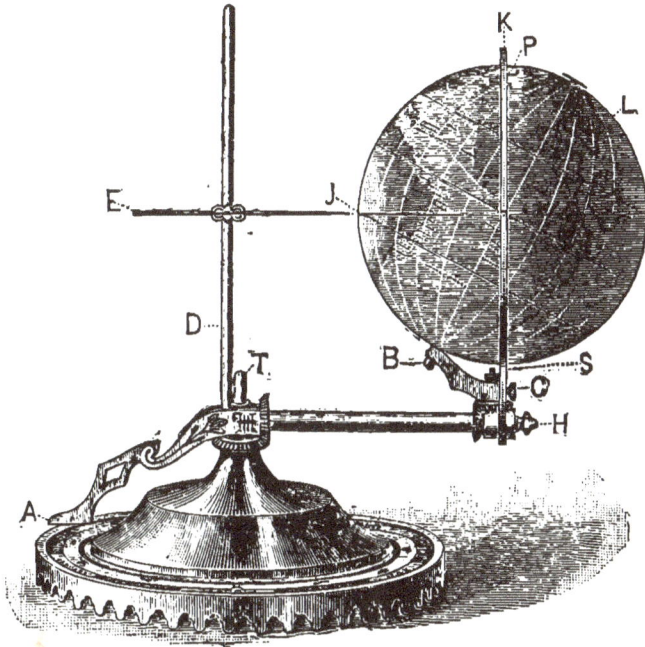

Directions for setting up the Tellurian.

Place the round base in such a position that Dec. 21st will be at your left hand and toward the south.

Adjust the globe as shown in the cut, move the geared arm until the calendar index A is exactly over Dec. 21st.

Adjust the short arm B C so that the stud underneath will drop into the hole in the gear below. Tighten thumb screw C. This movement places the earth in true position, from which it cannot vary. Screw in the sun wire D, and adjust the pointer E so that the point J will indicate meridian 95° (which, on our eight-inch globe, runs north through the United States) at the point where Tropic of Capricorn, the Plane of the Ecliptic and the meridian intersect. The tellurian is now *rectified for Dec. 21st* or Winter Solstice.

Adjust the day-circle K by sliding the thimble over the end of the horizontal geared arm at H.

Description.

The Sun is represented by the post D, with its horizontal moveable wire E ; the latter illustrates the central vertical ray of heat, or a straight line from the sun's to the earth's center.

The Sun is not a mere speck of light, but it is so far distant that its comparative bulk is out of the question. A tellurian constructed upon true astronomical proportions would be a huge affair, requiring a wide expanse of plain for a class-room. Given an 8-inch globe for our earth, our 2-inch moon would describe its orbit at 20 feet, while our sun would need to be a sphere 72 feet diameter at a distance of a mile and a half.

E J represents the vertical rays and maximum volume of heat bestowed upon the earth. If you move the solar index E up the post D and project it near the earth, you will see that it would, if *too near* the map, describe just such lines as now encircle the earth.

These Parallels of Latitude are numbered from the

Equator north to the North Pole, and south to the South Pole.

The greatest of these circles, the Equator, divides the map of the earth into northern and southern hemispheres.

Revolve the globe rapidly and describe these parallels. Adjust Solar index as in the cut and let it cover the exact point where the meridian of Greenwich crosses the equator, observing —

1. Parallels of Latitude encircle the earth and are used to find distances and measurements east and west or Longitude. These distances are measured from a line crossing the parallels and extending from pole to pole, called the Meridian of Greenwich or Prime Meridian.

2. Meridians are lines extending north and southward and are numbered by degrees, beginning at the Equator, therefore —

The latitude of a place is its distance in degrees north or south from the equator. The longitude of a place is its distance east or west of the Prime Meridian.

Push the solar index E down to the Tropic of Capricorn and let it rest. Turn the globe on its axis until the index covers the point where meridian 95 crosses the Tropic of Cancer, and let it rest.

The Tellurian now illustrates the position of the earth Dec. 21st, when it is noon on meridian 95. The Sun's vertical rays are at Capricorn, south of the equator; the Arctic region around the north pole is enshrouded in twilight and darkness; the Antarctic region near the south pole is bright with perpetual day. It is winter north, summer south of the equator.

Apparent Motion.

The unwearied sun, from day to day,
Does his Creator's power display. — *Addison.*

To the human eye the sun at intervals makes its appearance upon the eastern horizon, pursues a course upward until noon-day, thence downward to night-fall at the west, silently sinking to rest.

Its light is sublime, its silence awful, the calm dignity of its course is majestic !

The place of rising and setting varies, for we have observed that in summer it rises far enough north of east as to throw its rays through our northern windows; setting to the north of west, while at noon its path is traced nearly in the zenith.   In the winter its course is far to the south.   Late it rises from a royal couch far away on the southeastern horizon; so far and so late, that to us living near the 40th parallel, its daily visit is brief.   Early in the afternoon it retires, having made so short a journey through the sky that its rays give but meagre comfort to the frost-bound denizens of the north temperate zone.

Sun's diurnal course.

And when the evening shades prevail,
The moon takes up the wonderous tale,
And nightly to the listening earth,
Repeats the story of her birth. — *Addison.*

Moon's diurnal course.

The lunar journey is even more various and puzzling than the sun's, for the moon, apparently, suits her convenience as to place of rising and setting, as well as to time; rising, sometimes, close on the heels of her lord's departure, as if in hot pursuit ; at others, so tardy in her course as to find her light extinguished by his rays, belated in the morning sky.   Her moods are changeable, occasionally averting her face with quiet indifference, again beaming over the eastern horizon with a round, full, silvery stare; often disappearing entirely for several nights. Compared to the sun's constancy, the moon's conduct is strange to the extreme, whimsical and coquettish ever, yet always resorting to the same old tricks.

Stars of electric brilliancy rise and set, while others, less bright, are fixed; though each night their groups seem to have moved slightly westward. *Apparent course of the stars.*

The uniformity of the movements of these heavenly bodies — sun, moon, and stars — is deceptive. It is not easy to believe that the sun stands still, while the earth, whirling on its own axis at a rapid rate, performs a yearly journey through space, passing completely around the sun in an orbit of immeasurable magnitude.

> What though in solemn silence all,
> Move 'round this dark terrestrial ball?

is the poet's expression of apparent motion. We are sure the sun moves daily through the heavens, for we have all our lives watched its coming and going. The features of our landscapes have become as familiar to us as the sun's course. Are not we, our home, our neighbors, our hills, plains, valleys, the winding river which *Familiar aspects.* forever runs toward the sea — are not these firm and immovable? The forest, whose darkening shades nerved us to quicker steps as we hurried home from school, the sun going down, still casts a long wintry shadow, or resounds with the songs of birds in summer. Later in life we sped through its depths behind jingling bells, our boisterous shouts waking lonely echoes, and startling the fox, squirrel, and rabbit.

2

We had *seen the sun set* and were glad, because we preferred the superb great white moon riding high in the heavens, and illumining our wintry landscape.

Visit the home of an aged resident and hear him recount the old ways, describe old landmarks. If a house has been lost by fire or flood does not the site remain? If the forest has succumbed to the axe of progress does not the site remain, even though covered by a thriving city? Have the hills been removed and the river stopped in its course? Yes, the fixed features of nature remain in the same relation to each other; your houses, farms, prairies, towns, counties, and states are always the same. Ascend the White Mountains and you will ever catch a glimpse of the sea; descend the cañon and see the sun for a brief space at midday; take ship for foreign lands, sailing whither you will, when you return all the facts of nature are the same.

Man accomplishes much, but he cannot change the actual geography of states; he can simply change names and titles. The sun, moon, and stars *appear* to rise and set; yet the fact is that the sun is at rest, and you, with your house, farm, landscape, your journey round the earth, all are forever in motion at the rate of thousands of miles every hour, every moment.

Daily experience affords many illustrations of *apparent motion*. The facts of *real motion* are mathematical and cannot be gainsaid. Apparent motion is a phenomenon. Real motion is a noumenon.

Real Motion.

Modern astronomy has shattered the lying heresies of antiquity. The simple cosmogony which is based upon

Fixed features of the landscape.

human vision has been displaced by a grand mathematical science, prolonged study, and the telescope.
Only great efforts of the mind and a severe struggle
against the evidence of our senses can convince us that
the earth moves, instead of the sun. Not until the last
century did the truth penetrate a human mind, but since
that time great progress has been made. To-day the
astronomer, fortified with his scientific knowledge, gifted
with constructive imagination, closes his eyes for a moment and sees by mental vision the actual motions of
planets, satellites, systems so complex that a verbal
description is a mere mechanical form of speech.

The Sun is the source of light to our solar system. Sun.
Backward, through all the countless ages, aeons, cycles
of time, his rays have benefited worlds ; through un-
born futures he will continue to confer this boon. Eternally he shines ; illuminating any opaque object within
reach of solar rays, bestowing to planets the opposing
forces of light and darkness.

Planets are illumined according to size and distance Illuminated
from the sun. The earth receives light upon one-half its hemispheres.
spherical surface, that is, an astronomical division of the
earth in two equal hemispheres might be made, and one
would be always light, the other always dark. If the
earth, then, had no daily rotatory motion on its axis, one
geographical half of its surface would always be light :
day and night could then exist only by the movements
of the inhabitants of the earth, passing from one to the
other hemisphere. Revolving as it does on its axis, each
diurnal period includes an alternating condition of shine
and shadow which we call

Day and Night.

Illustrate with the tellurian : Turn the earth globe on its axis from west to east, observing a diurnal change. When, for example, Charleston, S. C., emerges from the night-shade, passing under the day-circle, it is sunrise at that place ; when the meridian of that place is opposite the solar index it is noon ; sunset at the opposite rim of the day-circle, etc.

The day-circle defines the limit, in space, of the illumined hemisphere, and enables us to understand that geographical limits of day and night are not fixed upon the earth, but are constantly moving in zones of black and white from east to west as the rotation of the earth conveys its inhabitants from west to east.

Cause of day and night.

This fact understood we discover that darkness is caused by an eclipse of the sun by the earth. An observer on the earth is conveyed to a position whence the sun is invisible on account of the earth's spherical form. At the equator the entire diameter of the earth is between the sun and a midnight observer, the latter, if he wished to look in the direction of the sun would simply see the earth at his feet.

What is light ?

The absolute negation called *space* or *ether* becomes illumined by a positive chemical force called the solar rays, which, mechanically reflected and tossed to and fro between the earth and its atmosphere, becomes *light*.

Light *absorbed* is lost. The sands of tropical deserts possess the dual power, to a remarkable degree, of alternately absorbing and reflecting light ; the hot noonday with its blinding glare being succeeded by a night of darkness so intense that moonlight affords little relief ; on a cloudy night darkness is absolute.

Possible uniform periods of day and night : Illustrate with the tellurian. Remove the day-circle K and the globe; readjust the globe by carefully screwing the projecting pole B on the post at S.

The earth is now mounted on a perpendicular axis. Replace the day-circle and rotate the earth, observing that people the world over would have equal periods of day and night. _Possibly equal day and night._

Rotate the arm, passing the calender index through the 12 months of the zodiac. At any time of the year, upon any parallel or meridian, day-light would endure 12 hours, darkness 12 hours. The sun would rise at 6 every morning, and set at 6: 12 hours later to an observer at any locality. Now, as a matter of fact, this is true only twice a year (except at the equator), at the equinoxes, March 21 and September 21. As we well know the periods of day and night not only vary in length at different seasons on the same meridian, but they differ at the same time on different parallels of a meridian.

On Dec. 21st the Dictator of Peru at Lima observes that the sun rises at 6. The President of the United States, at Washington, also an early riser, gains two hours more sleep, rising with the sun at 8. Both these rulers dine simultaneously if at noon, and when the Dictator at Lima observes his sunset at 6 P. M. the President remarks that it is the shortest day of the year, and the sunset hour is 4 P. M.

People entirely ignorant of natural causes have observed that in this part of the world the day is longest in June and shortest in December, while the nights are the reverse.

Remove the day-circle, restore the globe to its true position, and again adjust the day-circle.

We know that the longest day must correspond to the shortest night, from the fact that the sum of both must always be the same, viz., 24 hours, or the period of one diurnal rotation of the earth on its axis. We know that

Increase of daytime. the length of day gradually increases from Dec. 21st to June 21st when it reaches its greatest limit; it then decreases slowly, a few minutes each day, until Dec. 21st. Of course the changes of day periods are the reverse of those of night. If a resident of Chicago were to visit Buenos Ayres he would find the same relative changes, with the important exception that the seasons would be reversed ; June 21st marking their midwinter height. Journeying farther from the equator he would find greater degrees of variation as he approached the poles ; upon returning to Quito he would notice that day and night were equal the year round.

Illustrate with the tellurian, calendar index at Dec. 21st: The day circle divides the earth into two distinct hemispheres. For every day of 8 hours in our latitudes there must be a night of 16 hours. For every day of 8 hours there must be a day of 16 hours at the same latitude south of the equator. All days are 12 hours long at the equator.

Rotate the arm, calendar index at March 20th; turn the globe on its axis:

Equal day and night. Days have increased in length as the poles approach the day-circle until, at this equinox, they divide the period of diurnal time equally with the nights. The north frigid zone has emerged from its long wintry night, the south frigid zone has retreated from its long summer day. Days north of the equator have increased in length while south of the equator nights have increased. It is spring at Chicago, autumn at Buenos Ayres.

Bring the calendar index to June 20th.

Now we observe a marked change. If it were possible to reach the north pole a strange sight would excite the wonder of the explorer. Arriving there on the morning of March 18th, day would be heralded by a *Polar day.* luminous sky, but the sun would not rise ; creeping slowly around the horizon, approaching nearer and nearer its edge, until March 21st, when its small disk would show itself continuously above, still creeping round and round daily in spiral circles, until June 20th. On that day the sun would describe a circling course 23½° above the horizon. The explorer would seem to stand still, no points of compass to guide him, all meridians converging under his feet. All changes, variations, and possible alternations of day and night and time would have no significance to this amazed admirer of perpetual day.

Pass the calendar index slowly to Sept. 21st.

In consequence of this motion of the earth in its orbit one-half the area of the north frigid zone is removed from the sun's light. The explorer now observes that the sun, having slowly retraced its diurnal course in winding spirals down to the horizon, disappears for a long absence of 6 months.

But to an observer at the south pole on the morning of Sept. 22d, the sun would appear above the horizon in exactly the same manner as at the north pole 6 months previously.

From June 20th to Sept. 21st day decreases over the *Decrease of daytime.*

northern hemisphere and must, therefore, increase over the southern, while nights are in inverse order, until, at this autumnal equinox, as at the vernal equinox, day and night are equal.

Coming to our point of departure Dec. 21st, we find, upon our arrival at the winter solstice, that the entire arctic circle has passed beyond the limits of solar light, and for a few days there is no sunrise upon that parallel.

Increase of night time.

We have studied the four cardinal positions of our earth in its relation to the sun, viz., the position of the solstices and the equinoxes, about June 21st and Dec. 21st, March 21st and Sept. 21st, respectively.

We now learn why the tropics and polar circles are distinguished from the other parallels upon the earth's surface : for the former are the farthest parallels from the equator, which are illuminated by a vertical sun during the year ; and the latter are the farthest parallels from the poles, which pass wholly out of, or wholly into, the sun's light.

Variation of day and night on different parallels at one time.

When a parallel lies partly in the illuminated hemisphere and partly in the unilluminated, or is divided by the day-circle, it has a day and a night every 24 hours, or during every revolution of the earth upon its axis. When it lies wholly in one of these hemispheres, its day or its night continues until it is again divided by the day-circle. Now, as we have seen, parallels of the frigid zones are the only parallels, which, at certain times in the course of a year, are divided, and at other times are not divided, by the day-circle. These parallels are, therefore, subject to a greater variety of day and night, as regards length, than occurs within the temperate and torrid zones. They have, indeed, four distinctly marked periods during the year ; these periods varying in length according to the distance of the given parallel from the pole.

Naming these periods in the most convenient way, they are, 1st, a period of continuous day, during which the parallel is wholly in the illuminated hemisphere ; 2dly a period of alternate day and night, during which the parallel is partly in the illuminated and partly in the unilluminated hemisphere ; 3dly, a period of continuous night, during which the parallel is wholly in the unilluminated hemisphere ; and 4thly, a second period of alternate day and night, during which the parallel is again partly in the illuminated and partly in the unilluminated hemisphere.

The middle of a period of continuous day, for either the northern or the southern hemisphere, is at the summer solstice for the corresponding hemisphere. The middle of the succeeding period of alternate day and night is at the autumnal equinox. The middle of a period of continuous night is at the winter solstice. The middle of the succeeding period of alternate day and night is at the vernal equinox.

As an instance of this variety of day and night within the frigid zones, let us see how it is exhibited at Spitzbergen during the year. Day gradually increases in length, from a momentary glimpse of the sun on Feb. 21st to 12 hours on March 21st; then to 24 on April 21st, when it remains continuous until Aug. 21st; it then alternates with night, decreasing from 24 hours to 12 on Sept. 21st, and to a parting glimpse of the sun on Oct. 21st; when a continuous night of four months succeeds.

Farther south, as in the southern part of Nova Zembla, we should find a continuous day and night of about six weeks each, and periods of alternate day and night of twenty weeks each. Farther north, on the contrary, we should find that the periods of alternate day and night are shorter, until, at the poles, they cease altogether, and the two periods of continuous day and of continuous night, each 6 months in length, compose the year.

The greatest length of day within the torrid zone is about $13\frac{1}{2}$ hours, this length occurring upon either tropic. The greatest length of day within the temperate zones obtains upon the polar circle, where it is 24 hours. The length of any given day subtracted from 24 hours gives the length of the night, and *vice versa.*

Owing to the variable rate at which the sun moves between the tropics, the relative length of day and night, *at the same place,* also changes at a variable rate. Thus this change proceeds the slowest at the times of the solstices, and the fastest at the times of the equinoxes. The reason of this is to be found in the varying inclination of the sun's motion along the ecliptic to the equator. Thus, for some time before and after the solstices, the sun is describing an arc of the ecliptic (40 or 50 degrees in length) which is nearly parallel to the equator on both sides of the solstitial points ; consequently the change in the sun's declination during this period is very small.

Variation of day and night at the same place.
During this period, therefore, the sun describes diurnal circles which nearly coincide, and the length of day is nearly constant. On the other hand, about the time of the equinoxes, the sun's course has the greatest inclination to the equator ; and, therefore, the change in the length of day is most rapid. The following is an esti-

mate of the rate at which its declination increases from
the time of the vernal equinox.

From March 21st to April 21st the sun moves northward about 10°; from April 21st to May 21st about 9°;
from May 21st to June 21st about 4°. The same cause
also affects to the same extent, the rate at which the sun
moves to and from the zenith between successive noon-
days, and also determines the rate at which it advances
along the horizon between successive sunrises and sun-
sets. It follows, from the equality in the rate at which
the sun is changing in declination at dates equally removed from either solstice, that the earth's surface is
illuminated precisely in the same manner on any pair of
such days.

Hence, for every day in the year except those two Annual sum
of day time
which date at the solstice, or the longest and shortest and night
time.
days, there corresponds another day of the year equal to
it in length ; also, at dates equally distant from either
equinox, the sum of the length of these two days or of
the two nights, must equal 24 hours. Every place upon
the earth's surface has six months day and six months Six months
day and
night during the year. At the poles, as we have seen, night.
the year is divided into a day and night of six months
each ; at the equator every 24 hours is divided equally
into a day and night. At places between these two
positions the sum total of the length of day and night is
known from this fact, viz., for every day shorter than 12
hours, during the year, there corresponds one as much
longer ; and the same is true of night, making the average length of each 12 hours for the year ; in which time
the total length of each must sum up to 6 months, as at
the equator.

The varied changes in the length of day and night
may be also studied to advantage by examining the diurnal course of the sun with reference to the horizon of a
place during the year.
The daily motion of the earth upon its axis from west
to east causes an apparent motion of the sun across the
sky from east to west. Now, it is evident that the
apparent motion must be at the same rate as the real;
that is, the sun moves both above and below the horizon
at the rate of 15° every hour, or 1° every 4 minutes.
The motion is performed in a great circle (viz., the celestial equator) when the sun is vertical at the equator, or
upon March 21st and Sept. 21st; at all other times. the
diurnal circles described by the sun are small circles.
An observer at the north pole upon March 21st or Sept.
21st would be able to follow the sun's course completely
round the celestial equator, which, at the pole, coincides

Polar
sunrise.

with the horizon (see Fig. 7). In fact, at the pole. the
sun is rising upon the former date, and setting upon the
latter. If he continued to observe this course for three
months subsequently to March 21st, he would find that
the diurnal circles described by the sun gradually decreased in size. just as parallels decrease from the equator to the tropics. The sun's diurnal course is not, however, an exact circle; since, while describing this course,
it is changing in declination; it follows, in fact, the
direction of a spiral.
The sun ascends about 10° above a polar horizon during the first month of the long polar day, 9° during the
second, and 4° during the third; attaining a final distance of $23\frac{1}{2}$° above it, after which it descends at a corresponding rate.

This motion of the sun from and to the horizon is identical with its change in northern declination.

Description of the Sun.

The sun differs utterly from the other stars of our solar system. It is like none, and none can be compared to it. In physical and astronomical attributes it is peerless. Neither planets, satellites, asteroids, nor comets can approach it in importance. Its immense size, physical constitution, incomparable chemical properties, and eternal durance warrant its assignment to sovereign and isolated rank. Large enough to furnish a home for all the heavenly bodies that revolve around it ; six hundred times greater than all the planets of its system with their satellites and asteroids ; it is one million three hundred thousand times larger than our earth. The circumnavigation of the earth requires three years ; to sail around the sun would take 300 years. That vast solar globe is a molten mass, emitting perpetual fires ; a peculiarity not shared by any other member of our stellar world ; on the contrary, they are neither hot nor luminous of themselves, and if the sun were not, they would be plunged into eternal darkness, doomed to everlasting cold.

The light and heat of the sun are changeless, never losing their power. The sun is a fixed center of a group of stars revolving in their orbits around it at various distances. These planets have their moons which revolve around them as they perform their orbital relation to the sun. This fact seems to be the fundamental mystery of creation. We hastily imagine we know why it is so since Newton's discovery of the law of attraction,

The peerless sun.

A ball of fire.

Orbits.

but we mistake a word for a thing. Newton was careful to say that he proposed a *name* for a *phenomenon wholly* **Attraction.** *inexplicable* in itself, and of which we know only the external manifestation ; that is to say the mathematical law.

We know that bodies approach each other in the ratio of their masses, and in inverse ratio of the square of their distances ; but *why* they do so we do not know and probably never shall during our terrestrial existence. It is the cherished hope of the devout astronomer that a future and higher state of being may enable him to know all mysteries of creation as they are known to their Creator.   As it is we know only the outward signs **A universal** of that universal agency which binds worlds in correla- **force.** tive ties ; which holds and controls not merely physical masses, but manifests itself throughout the entire domain of Nature, permeating and dominating all life ; uniting states, races, classes. interests.

Newton called it ·· attraction," superseding the word ·· vortex " of Descartes ; Kepler used the terms "electrization," "affection," "sympathy," "obedience." These terms simply describe a force without describing its origin.   Obedient to this irresistible force the earth revolves around the sun, blessed with its beneficent floods of light and heat: from these twin forces we get our climates, seasons, days, and nights.

Vertical and Oblique Rays.

Heat radiated by the sun is absorbed by the earth during day and radiated during night ; hence it follows that, other things being equal, a 12 hours night would radiate the heat absorbed in a 12 hours day.   Heat

radiated by the earth at night is not lost in indefinite
space on account of the atmosphere which surrounds
the earth, serving it in the capacity of a hot house,
retaining and distributing solar light and heat.   It is
obvious that a minute planet floating in ether at an
incalculable distance from the sun, would not retain by
absorption an appreciable amount of heat but for the
agency of this spherical shell of atmosphere which is Earth's
not peculiar to our globe alone, but is the palladium of atmosphere.
all planets.

It is obvious, also, that the spherical earth receives
most heat upon that portion of its surface whose plane
is nearly rectangular to the direction of the heat rays,
viz., the torrid zone.   Geographical zones derive their Why the
significance from the facts relating to the sun's declina- map is divided into
tion!   Tropical temperatures are constantly at a high zones.
degree, owing to vertical rays ; while our temperate
zones are so named because at those latitudes oblique
rays impinge upon the convex, are repulsed and scat-
tered.   The atmosphere takes them up and by gradual
diffusion creates an average for a belt whose extremes
are far apart.   The divisions of zones are arbitrary and
relate to the sun's declination rather than to temperatures, and such a mathematical arrangement would not
exist but for the earth's inclined axis.   As has been
shown in a previous paper, if the axis were perpendicular to the ecliptic the duration of light would be uniform
to all parallels from pole to pole.   Heat under the same
axial conditions would be absorbed and retained by the
earth's crust with uniform intensity on a parallel the
year round, varying in ratio of distance from equa-

torial central rays. These facts bring us to one of the most interesting and puzzling of natural phenomena, viz.:

The Inclination of the Axis.

Milton says in "Paradise Lost" that, before our first parents sinned, perpetual spring reigned on earth; but that as soon as Adam and Eve had eaten the forbidden fruit, angels with flaming swords were dispatched from heaven to bend the poles of the earth more than 20°.

Milton's Angels.

It is fortunate for us that the angelic power stopped at 23½°, or else our season would have been still more abrupt in change. In Mercury the inclination is pro-digious — not less than 70°. This planet leans on itself

Climate of Mercury.

as if about to fall. The Mercurian climates would be unendurable to us, but a moment's reflection assures us that the Creator of worlds can enact His Divine will upon all his works, and that the dwellers in each planet-ary home are adapted to their environment, otherwise they could not exist.

Fourier's notion.

Fourier declared it to be possible for man to exert a power sufficient to readjust the earth on an axis perpen-dicular to the plane of its annual orbit, and to restore the equality of the seasons. He neglected to mention one important point, viz., the mechanical process by which this power could be exercised. His theory re-sembles the act of a drowning man who thought to save himself, by seizing his own hair; and that other theorist who had his suspenders made very strong. that he might by a shrug of the shoulders, lift himself over wet places.

The axis of Venus is inclined 75°. Babinet describes as follows the effect of their great inclination on seasons : ".The planet which certainly presents the most re-

"markable climatological peculiarities is Venus, which in Babinet on Venus.
" bulk and distance from the sun is almost the exact
"counterpart of the earth. She turns very obliquely
"on herself. If we take the earth for a point of com-
"parison, the sun in summer comes almost above Cuba.
" The obliquity of Venus is so great that in summer
"the sun attains latitudes higher than those of Belgium.
" It follows from this that the two poles, subjected in
"turn to an almost vertical sun, which never sets (and
"this at intervals of four months, since the year is but
" eight months long), do not permit snow and ice to
"accumulate.   This   planet   has   no   temperate   zone;
"the torrid and frigid zones encroach upon each other
"and rule successively over the regions which in our
"world constitute the temperate zone.   Hence results
"constant conflict of the elements agreeably to what our
"observation has taught us as to the difficulty of seeing
"the continents of Venus across the veil of her atmos-
"phere, incessantly disturbed by the rapid variation of
"the height of the sun, of the duration of days, and the
"transports of air and moisture that render the solar rays
"twice as powerful as those that come to the earth."

Reasoning from the analogy of the conditions of our
earthly life, Jupiter would seem to be a very desirable
place of residence for those discontented souls who long
for what they cannot have.   The axis of Jupiter is in- Jupiter's climate.
clined but slightly, hence it has, like Saturn, a kind of
perennial spring, that is to say, a reception of solar light
and heat operating in equal proportions along the same
parallels.   With short days and nights of five hours
each, its year is equal to twelve of ours.   Owing to great
distance from the sun, days can be but faintly illumined.

Astronomers there must be enabled to see the most beautiful stars at midday. In compensation for short nights Jupiter has near about him four moons which supply a steady light ; his twilights are long, hence the light of his day and night must be nearly equal. Perfect equality of days and nights and of seasons on all his parallels; so that the discontented Jovian needs but to change his latitude to find a season befitting his various requirements, desires, and whims.

Results of inclined axis. The inclination of the axis of a planet is the test of its condition as an abode of life in any form. In the study of our subjects concerning light and heat the inclination of the earth's axis is easily seen to be the cause of all variations of duration and degree in our reception of these two primal elements of life. We have observed that during an orbital revolution of the earth around the sun in a year, the poles vary in their angular relation to vertical rays, though always pointing in the same direction — toward the north star. This is called the "parallelism of the poles." This is a mystery to most pupils, for they are apt to wonder why we do not call the axis perpendicular and every thing else inclined. In their minds it is a mere matter of names, and if you try to explain by saying that the axis is inclined to the plane of the earth's orbit, they are worse off than ever ; for you have but added more names: and *names* are nothing without an understanding of the *thing* which is back of the name. Object lessons assist the imagination, making it not only active but constructive.

If you have made them familiar with the few simple natural facts of daily observations as herein described, such as these — that the sun is at rest and the earth

moves; that heat may be received alike at different in-
clinations, but *not* retained where the rays fall obliquely;
that heat absorbed during the day is lost by radiation
during the night, you have laid a foundation for some
beautiful illustrations.   When you tell them that the
earth's axis is *inclined*, if they are bright they ask —

" Inclined to *what ?* "

" The Plane of the Ecliptic ! "

" What is that ? "

If you are not familiar with the theory of the eclip- Plane of the
tic, do not say anything about it until you are; it is   ecliptic.
worse than Greek roots and French verbs on a slight
acquaintance.   Turn the earth or globe on its axis and
observe that there is a line which crosses the equator at
the prime meridian, runs southward to the Tropic of
Capricorn, upward across equator again to Cancer.   The
equator and parallels are always visible while the globe
is revolving, but *this* line wobbles about in a zig-zag
manner, and is invisble if the globe moves rapidly.
Bring this erratic line to a horizontal level and it is in
line with the solar index E and J.   The index and the
line now represent the Plane of the Ecliptic, or an imaginary line drawn from the sun's to the earth's centers.

Such a line would always touch the earth's surface, as The Plane
the solar index might, somewhere in the tropics, and   illustrated.
to that fact is due the division, by the early astronomers, of the earth's surface into geographical zones.
Rotate the arm and explain that the solar index points
at the ecliptic   But this explains only the geographical ecliptic marked on the map for convenience.
Rotate the arm entirely round the sun and explain that
the ecliptic is the level of the earth's orbit.   Are you

not as far from the real fact as ever ? For if you turn
the globe on its axis in the slightest degree, your line of
the ecliptic goes off its balance !

Bend the upper half of the day circle K down to a
level with J, and rotate the arm backward and forward.
You can by this means describe the fact that the true
ecliptic exists outside and away from the earth's surface, and you have gained a point. This experiment
proves that the Plane of the Ecliptic is parallel to or
level with the earth's orbit.

Tip up the tellurian and rotate the arm. The effect is
the same.

If your pupils whirl on their toes, without moving
out of their tracks, they will describe the earth's daily
rotation on its axis. If they run around a post they
describe an orbit. If they both whirl and run they
describe both motions of the earth in a year.

If you stand in the center of a circle of pupils and
pass a cord from one to another, around the circle,
holding the end yourself, the cord will, if level, illustrate the Plane of the Ecliptic. You are the sun, one-
half above, one-half below, the plane. Each of the
pupils illustrates the earth in one of its positions. The
whole circle is the orbit. Take a blackboard pointer
and explain as you pass its farther end round the circle,
that the earth always moves on that level; then raise
and lower your stick as you describe the moon's orbit
once a month round the earth, one-half the time above
the ecliptic, the other half below it.

Heights
from sea
level.

A house upon a mountain top is thirty feet high.
High above what ? The mountain. How high is the
mountain ? Well, the surveyors were around a year or

two ago and they said the top of the mountain was 2,000
feet above the valley where the water of the river is
made level by a dam. But how high is the dam ? The
surveyors said that there were ten feet fall between the
reservoir and the factories below. How high are the
factories? 40 feet from the river where the water is
returned to its natural channel. How high is the river
at the factories? 50 feet above the sea level. Now
we have reached our destination! The mountain, then,
is 2,100 feet above sea level.

The early surveyors of the skies found a level for Astronomi-
their base line. It is the Plane of the Ecliptic — the path
in which the earth always moves in its annual pilgrim-
age round the sun. Describe a circle on the blackboard, one foot in diameter. A few feet distant describe one a trifle smaller. Draw a horizontal mark
through the middle of both, a *continuous* line. The
larger circle is the sun, the smaller the earth. The
line is the ecliptic. Erase the earth and describe it the
other side of the sun. It has passed from winter to
summer but it has not left the line. Describe the same
by rotating the arm. The earth does not jump up and
out of the Plane of the Ecliptic as the moon likes to do;
as a flying fish might dart up and out of the sea, then
back again to the depths below the sea level. Our re-
spectable and dignified planet moves forever along the
same level, one-half above, the other half below, very
much as the hull of a large ocean steamship moves over
water, one-half submerged. The theory of the ecliptic
once made plain, we pass on to an acquaintance with the
Zodiac.

The ancient astronomers discovered that the sun

appeared to have passed among certain groups of fixed stars. To these groups they gave the names now used. Signs of the zodiac. These constellations occupy space in a belt of the heavens parallel to the plane of the ecliptic and extending 8 degrees above and below it. Within the limits of this Zodiacal Belt all the principal planets of our solar system have their orbits. But the earth is the only one that sticks to the dead level of the plane itself. They divided the belt into 12 portions of space to which they gave the Signs of the Zodiac.

Retaining this arrangement, modern astronomers say that on March 20th the sun enters the first point of Aries. Illustrate by bringing the solar index to March 20th. Thus, while the sun is in one sign, the earth, as seen from the sun, is in the opposite one.

Rotate the arm and observe that the earth is above one sign when the sun is over a sign on the opposite side of the circle. Illustrate better by allowing the pupil to stand on the night side of the globe. Elevate the solar index to the top of post D and ask them to *squint* along the direction of the solar index to a distant window. Rotate the arm a little and they may change their positions and discover another window, etc. Call one window Leo, and let them pass around to the day side and look in the opposite direction. The earth is in Aquarius, where the sun will be in six months. Rotate the arm to prove it.

You can, if you choose, easily construct a good Zodiacal Belt; take strips of pasteboard a few inches wide and by fastening the ends together, make a circle four feet in diameter. Divide the inner surface into 12 parts and name them. Make a distinct mark through the middle of your belt, entirely around. Now arrange your circle on a level at the right height and place the tellurian in the center of it. If you have been careful, the solar index, restored to its proper place on the post, will point constantly at the dividing line on your belt, as you rotate the arm. That line is our old friend the Plane of the Ecliptic, and the 12 spaces are the Signs of the Zodiac.

> " The Ram and Bull lead off the line;
> Next Twins, and Crab, and Lion shines;
>   The Virgin and the Scales.
> Scorpio and Archer next are due,
> The Goat and Water-bearer too;
>   And Fish with glittering scales."

The appearance of the sun and moon in these signs at regular intervals of years and months proved to the early observers that the sun described an orbit around the earth, or else the earth had an orbit around the sun ; whichever it might be, there was a *level* — the Plane of the Ecliptic.

Having found a base, we know that the earth is inclined to that level because of the variation in the sun's apparent altitude.

Zenith is a point of the farthest extremity of a line Zenith. extending from the center of the earth, through its crust, into space ; the zenith of any observer is directly overhead as he stands erect and looks upward.

Rectify the globe for Dec. 21st.

The solar index E points at the Tropic of Capricorn; a resident of one of the Gambier Islands sees the sun in his zenith at noon, Dec. 21st.

Rotate to March 20th.

A resident of Quito sees the sun in his zenith. Rotate to June 20th. The sun is in zenith, or exactly overhead at Havana, and appears nearly in the zenith as far north as Boston; but in fact is 20 degrees south of zenith; its great distance lessening the angle of rays to an observer on the 45th parallel. When the sun is in zenith at Gambier Islands, its vertical rays fall upon a zone north and south, 45 degrees in width. To us above the 40th parallel, the sun, Dec. 21st, *appears* far south at noon, while on June 20th it *appears* nearly overhead. Between these extremes of low and high sun is the apparent variation of sun's altitude.

High and low sun.

You may hear people remark in February that winter cannot last much longer because the sun is "high."

In the autumn we are warned of approaching winter by a "low sun." When the sun is in Capricornus it is "low," but rises higher every day until it is in Cancer. You will find in the almanacs little crescent-moons with the horns pointing upward or downward, signifying that the moon runs "high or low." Rotate the arm entirely round the sun, observing that a person must, if he wishes to keep the noon sun in his zenith, travel north from Cobija, South America.

High and low moon.

When should he start? About January 10th. At what season? *Our* midwinter and *his* midsummer.

Why not start Dec. 21st?

Let us see; rectify for Dec. 21st.

This is called the Winter Solstice, because the *sun stands* at the same altitude at noon for nearly a month.

Turn the arm backward until the Cardinal index covers Nov. 10th. Observe that the ecliptic line has nearly reached the Tropic of Capricorn. Owing to its

distance the sun's altitude at noon would appear to be
zenith at Gambier Islands and it would rise and set in
the same place for nearly a month.   Our Gambier
Islander would travel almost exactly eastward, during
that time, starting early in November.   He must needs
sail 4,000 miles in 60 days.   Describe his slow journey
under the solar index by slowly bringing the calendar
index to Dec. 20th.   We would read in the paper that
while he was melting under the vertical rays of midsummer sun we were burning our income in coal to keep
from freezing under the oblique rays which fall upon
our northern regions in midwinter.   To make matters
worse, we would notice that the improvement in the
angles was very slight, for the sun, to the northern observer, rises, culminates, and sets in the same little arc
day after day, until January.   An observer at the Arctic Circle would see the sun barely above his southern
horizon at noon, for many days prior to Dec. 21st.   On
that day the sun merely peeps over it and drops back
out of sight   Old Sol sends his regrets to the Esqui-
mau, pleading other engagements.   He is bound to give
our traveler a good hot noon, and has promised to stay
up all night with some friends at the Antarctic Circle.

By Jan. 10th, the tropical traveler lands on the western coast of northern Chili, and sleeps at Cobija.

Every 24 hours he has sailed eastward 67 miles,
reckoning from his horizon.   Each day his sunrise,
noon, and sunset have been described by the *apparent*
motion of the sun through an arc extending from east
to west, through zenith.   And his latitude has changed
but few degrees, because his journey up to this point
has been along a route where the noon sun is nearly at
the same altitude.

A lover of vertical rays.

Old Sol's bad habits.

His days have been about fourteen hours, his nights Our tropical traveler. ten hours long. At Cobija he finds his watch does not agree with meridian time. He has journeyed eastward 4,000 miles in sixty days, leaving home November 10th, 1885, arriving at Cobija as he supposed at 10 P. M. January 9, 1886, but upon correcting his watch he finds it to be January 10th, 2 A. M.

He has kept his diary every day and noted the fact that the sun rose earlier each day, and set earlier, but he has gained four hours by traveling toward sunrise. The captain of the ship bids him good-bye at Cobija, remarking that he will not change his watch as the four hours will be restored by the time he reaches home, sailing westward.

At sunrise, Jan. 10th, our friend telegraphs an acquaintance in London that he shall continue his experiments with vertical rays, therefore he shall travel northward to Quito. It takes six hours to get the dispatch to London, and as the sun rose at 6 at Cobija, should it not be 12 in London when the dispatch is received? Bring the Cardinal index to Jan. 10th. Rotate the globe until Cobija is at the western edge of the day circle.

Point the polar index to 6; rotate the globe slowly until Cobija is under the solar index.

Where is London?

Arriving at Quito March 20th, his watch tells him that it is noon, but it is nearly sunset in fact. He has lost over five hours by pursuing a northward course 1,400 miles. The journey of the tropical traveler reveals a few natural facts:

His discoveries.

1. Variation of time.
2. Variation of area of Vertical Rays.
3. Variable Reception of light and heat.

These phenomena are caused by the double motion of the earth; viz.: a daily rotation around its own axis, a yearly rotation around the sun, both motions being accomplished with one-half of the earth heated and illuminated by the sun.

To find where Vertical rays visit.

Rectify the tellurian for Dec. 21st. Bring the calendar index over the month or day designated. Rotate the earth and the solar index will cover the area of vertical rays. Practically, the area extends 23½ degrees in any direction from the point covered by the solar index.

Examples.

What countries receive vertical rays in Feb., May, Aug., Nov.?

Upon what parallels do vertical rays fall in June?

There can be no such a thing as unequal *distribution* of light and heat, for, as has been remarked before, the sun constantly, invariably distributes both impartially. We have seen that if our globe, the earth, were mounted on a perpendicular axis, all portions of its surface would receive light and, relatively to latitude, heat equally. Day and night would be equally divided the world over.

Rectify the tellurian for Dec. 21st, and rotate the globe, observing that while one-half of the earth receives the light of the sun, the opposite half does not. Rotate the arm to June 20th. The earth continues to receive the same amount of light and heat, but, by its changed position, the Arctic region is now brought into the light.

An observer upon a parallel would see the sun in its apparent course through the sky describe an arc of the

The tireless sun.

same degree of curvature as the parallel upon the globe. We find that the *reception* by the earth of light and heat varies according to the earth's position in relation to the sun.

Observe that the southern hemisphere is projected into constant day.

What is the duration of light at the Antarctic Circle? 24 hours.

At Cobija, South America?

Bring Cobija to the point of sunrise and set the polar index to 12.

Examples.   Rotate the globe on its axis and you find that when Cobija touches the point of sunset the index marks 2 + 12 = 14 hours.   Divide by 2 and you get 7 hours A. M. and 7 hours P. M.   Their sunrise must therefore occur at 12 — 7 = 5 o'clock.   Their sunset at 7 o'clock.

Bring Quito to sunrise and proceed as before.   The index gives a 12-hours day.   Try Cape Sable, Florida; you find a day of 10 hours — sunrise at 7, sunset at 5. Try San Francisco: nine hours of light and 15 hours of night.

Try Vancouver's Island: $7\frac{1}{2}$ hours of daylight and $16\frac{1}{2}$ of night.

Try the North Pole!

Continue these experiments, observing that, at the Solstices in Dec. and June, light and heat, though equally distributed by the sun, are variably received by the earth.   A northern winter is offset by a southern summer and *vice versa*.   At the equinoxes, March 20th and September 23d, heat and light are equally received on all portions of the earth's surface.

A Few Convenient Rules.

Given the length of day, subtract from 24 to find length of night; divide the hours of night by 2 to find sunrise; divide the hours of day by 2 to find sunset.

Given the hour of sunset, subtract from 12 to find sunrise.

Double the time of sunrise to find length of night.

Double the time of sunset to find length of day.

Twilight.

The daily morning and evening glow, before sunrise and after sunset, would not exist but for the gaseous envelope which surrounds the earth. Atmosphere is about 50 miles thick, catching the tangential rays of rising and setting sun, bending them downward toward the earth by refraction, and diffusing them by reflection. Without the aid of atmosphere solar light, striking the earth's surface, would be reflected and instantly lost in etheric space; in such a manner that, though to an observer on another planet the earth would look bright, the observer on the earth would gaze off into deep dark-ness even at midday; the sun appearing to resemble a full moon. No morning twilight would herald the sun's approach; no evening graduation would warn us of the coming night; but abrupt changes from one extreme to the other would be our lot. If by some astronomical calamity, the atmosphere should suddenly cease to exist, sound would terminate in silence, respiration become impossible, the frigid cold of space would seal the earth's crust in eternal frosts, rivers, lakes, and seas be swallowed up by universal absorption; the planet on which we dwell would become, in one brief moment, a voiceless, lifeless orb.

We have observed that one half the earth is always bathed in light; between the two hemispheres of day and night exists a narrow belt encircling the earth, about 18° wide, called *twilight.* The width of this belt is, theoretically, 18°, but this measurement denotes an average rather than a uniform rate. Twilight is increased by cold and decreased by heat; its duration also depends a good deal upon locality, varying 5° from equator to poles. The time required for the sunrise line to travel 18° depends not only upon the angle between the circle of declination which the sun is describing, and a given horizon, but also upon the size of this circle of declination. The only horizons for which the entire diurnal course of sunrise is at any time coincident with a vertical circle are the horizons of places upon the equator at the times of the equinoxes; the shortest twilight, therefore, is seen at the equator on March 21st and Sept. 21st, enduring only 1 hour and 20 minutes.

The higher the latitude the greater the obliquity of its horizon to the sun's course; it is plain, therefore, with the aid of the tellurian, why twilight varies from equator to pole; simply because an observer at the equator is carried, by the earth's motion, through the twilight belt with greater speed than he would be at high latitudes.

Illustrate with the tellurian :

The space between the day-circle K and the nightshade L is the twilight belt. Outside the Arctic and Antarctic regions there are two twilights each day. Rotate the globe and observe the increased obliquity of parallels, crossing the belt, as you pass from equator to poles.

Extent of twilight.

Shortest twilight.

Bring the calendar index to Dec. 21st:

The Arctic region at the time of winter solstice, has twilight and night for its alternatives ; twilight increasingly succeeded by daylight until the time of vernal equinox, March 21st.   At this time daylight and twilight are the alternatives — no night.   Day now increases upon Alterna-tions. twilight until the time of summer solstice, June 21st, when it is all day.   Continue the observations around to Dec. 21st. not forgetting to notice that the order of change and duration at the Antarctic region is the same, though in inverse order.

To find the duration of twilights :

Bring the given place to the eastern slope of the day-circle, if in winter, to the western if in summer, to find evening twilights ; set the polar index to 12, and rotate the globe until the place is under the edge of the night-shade.   The index will show the duration of both morning and evening twilight.

To illustrate : Bring Cape Farewell under the day-circle Dec. 20th.   Index 12.   Rotate the earth and find a long twilight of 6 hours.

The Sun's Declination.

We have used the term ''sun's altitude'' to express what is apparent from the earth.   The sun *appears* high or low according to the season.   Astronomers use the word ''declination'' to define the position, north and south of the equator, upon which the vertical ray, the Prime verti-cal ray of central ray, the maximum degree of heat falls. heat.

When this ray reaches the earth north of the equator, the sun is said to have a northern declination; when south of the equator, a southern declination.   We have seen that the greatest northern declination is at the Tropic of Cancer, June 21st; the greatest southern declination is at the Tropic of Capricorn, Dec. 21st; March 20th, and Sept. 23d, the sun has no declination.

Illustrate with the tellurian :

Bring the calendar index to March 21st. Adjust the solar index by sliding up or downward until its point J covers the equator. At the times of equinox the sun has no declination, its meridian altitude being reckoned north and south of that line, which is zero. Rotate the globe from west to east, observing that an observer upon any parallel of the earth's surface would see the sun rise, culminate, and set, describing an arc corresponding in degree to that upon which he stands, measured from edge to edge of the day-circle.

Increase of declination.
Rotate the globe rapidly, bringing the calendar index slowly to June 21st. When the globe stops, observe that the solar index now covers the Tropic of Cancer, and for each day during the intervening period of three months the sun has had a *circle* of *declination* and meridian altitude. When the sun is north of the equator, northern circles of declination have their longer arc above the horizon and shorter below it, therefore days are longer and nights shorter than 12 hours.

Increased variation of time at high latitudes.
The greater the distance from the equator the greater the difference between the length of the two arcs into which a circle of declination is divided by the horizon ; and, of course, the greater disparity of day and night.

The more oblique the sphere, or the higher the latitude of a place the more does its horizon differ in direction from that of the sun's course, hence the longer the distance measured upon the horizon which corresponds

to a change of a given number of degrees in declination. In high latitudes the sun rises and sets at a Rapid advance of
more rapid rate of advance along the horizon, and also points of
sunrise.
approaches more nearly to its northern and southern
points. The horizon whose inclination to the equator is
$23\frac{1}{2}°$ has the sun within its northern or southern point
at the respective times of the solstices ; so that sunrise Sunrise at
the polar
and sunset must take place within every point of this circles.
horizon during the year, and one diurnal circle be
described above it at the summer solstice. This is the
case at places upon the polar circles, as may be seen
with the tellurian at June 21st, the arctic circle being
projected wholly within the day-circle.

At places within the polar circles the sun rises and
sets at every point upon their horizons during the year;
passing along the horizon at a more rapid rate the nearer
the place is to the pole, and reaching the north or the
south point *previously* to the time of a solstice. The *time*
previously depends upon the distance of the given place
from the polar circle, being greater with an increase of
latitude. As the latitude increases, the greater, there- Diurnal
circles near
fore, must be the number of entire diurnal circles which the poles.
the sun describes continuously above the horizon during
the year ; and this agrees also with the increase in the
continuance of day in this direction. The angle between
these circles and the horizon diminishes towards the
poles, until, at the poles, a difference of direction between the two ceases altogether, and the sun moves
either in the horizon, or in a direction parallel to the
horizon.

The rate at which the sun's diurnal circles ascend
above the horizon, or descend towards it, within the

4

polar regions (not now including the poles), depends upon the position of the sun within the ecliptic when describing them. If the day is two months or less, the sun is, during that time, describing an arc of the ecliptic nearly parallel to the equator, and moves northward or southward, therefore, at its slowest rate. When the day is a longer one, the sun leaves and returns to the horizon at its most rapid rate, or nearly so, during the former and the latter portion of it.

Sun's course at Spitzbergen.

As an instance of the variety which the sun's diurnal circuits exhibit in the frigid zones, let us follow the sun's course at Spitzbergen for a year, beginning with the dawn succeeding a period of continuous night.

The sun appears in the southern point of the horizon upon Feb. 21st, and immediately sets without an intervening course. The next day it describes a small arc,

Arctic day.

the next a longer one, and so on, until it rises in the east and sets in the west, having moved, both in rising and in setting, through a quarter of the horizon in coming to these equinoctial points. It now culminates 10° above the horizon. Subsequently to March 21st it describes arcs increasing in length until April 21st, when its whole circuit is brought above the horizon, and so remains, rising higher and higher, until the time of the summer solstice. At this time the sun comes to the meridian of Spitzbergen at a distance of 35° above the south point of the horizon, and 13° above the north; the difference between these numbers (or 22°) showing the obliquity of its diurnal circles to the horizon of Spitzbergen. After this date the sun begins slowly to descend, until it sinks below the northern point of the horizon for a moment on August 21st; after which it describes arcs gradually decreasing in length until it has left the visible heavens altogether upon Oct. 21st.

The zenith distance of the sun at noon for any given place north of the equator is equal to the difference between the latitude of the place and the sun's declination, if the latter is north declination ; or to their sum, if it is south declination. The meridian altitude of the sun is always equal to the difference between its zenith distance and 90°. The reason for these rules should be found by the learner; and he should also derive a corresponding rule for places in the southern hemisphere. Thus, when vertical at the equator, the sun culminates 20° from the zenith, or 70° above the horizon, at the latitude of 20° N. or S., towards the southern point of the horizon at the former latitude, and northern at the latter. When vertical at the Tropic of Capricorn, its zenith distance is $23\frac{1}{2}$° at the equator, $33\frac{1}{2}$° upon the 10th northern parallel, $67\frac{1}{2}$° at the south pole, and so on. *Meridian altitudes.*

Whenever the sun is vertical at a latitude north of a given place, it culminates at a point of the meridian north of the horizon of that place ; whenever it is vertical at a latitude south of a given place, it culminates south of the horizon of that place. It therefore culminates both upon the north and the south sides of the zenith at places within the torrid zones, but only upon one side of the zenith at places within either temperate zone ; namely, always towards the southern point of the horizon north of the Tropic of Cancer and always towards the northern point of the horizon south of the Tropic of Capricorn. Beyond the torrid zone the sun culminates at its extreme zenith distances at the time of the solstices ; occupying its nearest position to the zenith *Sun at noon north of zenith.* *South of zenith.* *Either side of zenith.*

at noonday of the summer solstice and its farthest position at the winter solstice; June 21st, north of the
equator; Dec. 21st, south of the equator.

Change of Seasons.

Variation and change of seasons. The meridian altitude of the sun at a given time denotes the season. The difference between vertical and oblique rays indicates the *difference of seasons*. The process of differentiation causes *change of seasons*.

At the equator there is but one season, perpetual summer; though a given locality may have its climate which is due to local causes. Away from the equator seasons change, even within the tropics, the degree of variation always being in ratio of distance from the equator. If All climates and no seasons. the earth had no orbital motion, there would be different climates for different parallels, but no change of seasons; but the orbital motion constantly changes the angle of the plane of a parallel circle and the vertical Causes of seasons. rays; hence we may say change of seasons is caused by differing durations of day and night; or, which is equivalent, the amount of heat daily absorbed and retained by the earth.

With the tellurian rectified for Dec. 21st, we observe that the declination of the sun is at the extreme southern point. Bearing in mind that all natural conditions are reversed for opposite directions from the equator, we will confine our attention to the relations of seasons to our northern latitudes. At this time of winter solstice, the earth is frost-bound. Since Sept. 21st the days have decreased and the nights increased in length; A northern winter. heat absorbed during the summer has been lost by radiation; cold has accumulated, and winter reigns. Very

oblique rays afford but little heat, too little to counter-
act the long cold nights, and all the veins of vegetable
growth are sealed, the sinews of the soil paralyzed by
the icy grasp.

The average man earns barely enough during the year
to keep his family comfortable, for a long winter calls
for increased expenditure for clothing, food, and fuel to
sustain health. *Its hardships.*

Our rigorous wintry climate is a terror to the feeble,
but to those who have and know how to maintain good
health this season has its compensations in facilities for
recreation, repose, study, and meditation. Civilization
depends in a great degree upon long nights. High
grades of intellectual attainment are commonly found in
high latitudes. Nature is silent with sullen impotence,
save the fierce clatter of hail, blinding blasts of snow,
and the terrific howls of the Manitoba blizzard. *Its advantages.*

It may be asked why, at March 21st, when day and
night are equal in all places, the average temperature
is not the same as at Sept. 21st when like conditions
prevail ? *Spring.*

The explanation is found in the fact that during a
northern winter the crust of the earth accumulates so
much cold that at the vernal equinox the gain in solar
heat has been lost in the reduction of cold. It requires
as much heat to melt a sheet of ice one inch thick as
would suffice to raise the temperature of 800 cubic feet
of air from zero to 100. One can readily perceive why,
in our latitudes, the climate is not so warm at vernal
as it is at autumnal equinox, because at Sept. 21st the
accumulated heat of the summer has not yet been lost
by radiation. Therefore, with the axiom that a 12- *Reasons why our climate is colder March 30th than September 20th.*

hours' night will radiate the heat of a 12-hours' day, we learn that where this equality of day and night does not exist for any length of time, the climate depends upon what the soil has accumulated as well as upon what it is daily receiving and dispensing.

To thoroughly understand the difference in degrees of reception and accumulation of heat let us experiment further.

Bring the calendar index to June 21st :

Summer.   The northern hemisphere is projected farther into the illumined space, toward the sun, where light and heat always prevail. This change of position has come gradu- ally since the advent of spring. Days being longer we must have received more heat than our nights alone would radiate ; what becomes of the surplus? It is absorbed in the process of liquefaction and evaporation. The sun *thaws* or melts snow, frost, and ice into water, which evaporates and flows away to the sea. We do not experience genuine summer heat until after we have passed the period of vertical rays and are well on our way to a position of oblique rays. Hence, we accumu- late heat during the shorter days which follow the time of solstice, June 21st, provided the days are longer than 12 hours. In July, August, and September we have our highest temperature, for then, and not till then, is our soil in a condition fit for accumulation of heat.

Earth's cen- tral heat.   An important influence which tends to ameliorate the condition of a cold season is the internal heat of the earth. All planets have a temperature of their own, independently of the sun. They were originally in a liquid state produced by heat, and have become fit for organized life by cooling to a solid condition. This individual heat is still preserved in the center of the earth. At any considerable depth below sea level this heat becomes perceptible.

As a result of this brief study of the seasons, we find :

1. Distribution of heat is always equal.

2. The sun emits only vertical rays, the *obliquity* appertaining to the angle *upon* which they fall.

3. *Reception* of heat is equal at all times and places during day, but *accumulation* of heat depends upon the relative preponderance of the day over the duration of its corresponding night.

4. Vertical rays are positive only at the tropics ; outside that zone they are negative through the process of reducing accumulated cold.

5. Reception and accumulation of heat depend more directly upon the relative duration of day and night than upon vertical rays.

Climates.

The tellurian illustrates, as well as any instrument can, the important and interesting phenomena of climates, and their relations to seasons.

Seasons are belts of mean temperature encircling the earth in zones parallel to the equator ; and the name, *summer* or *winter*, at a given place would apply to all places upon that parallel ; without regard to local variations of climate. When it is spring at New York it is spring at Madrid, Bokhara, and Pekin, but the climates of their cities are not alike. Difference between seasons and climates.

The exact classification possible to seasons is forbidden to climates ; the study of which is complex, embracing phenomena varied in character and little allied to each other. Modern geographers apply the term "climate" to designate the mean temperature of a place.

The world is indebted to Sir William Herschel and Alexander Von Humboldt for the rapid progress made in the science of climatology within the past 60 years. Humboldt published in 1817 his celebrated treatise on the isothermal lines, in which he showed that the decrease of heat with the increase of latitude takes place more slowly on the west coasts of the old world than on *Humboldt's researches.* the east of the new. He connected places having an average amount of temperature during the year by isothermal lines, the convex summits of which fell near the west coast of the old world, and their concave near the east coast of the new. By combining the decrease of temperature by increasing elevation with its decrease by increasing latitude, he represented the intersection of isothermal surfaces with a vertical plane cutting the surface of the earth along a meridian, and showed that if the examination of places of equal summer heat and equal winter cold were conducted in a similar manner by drawing isothermal and isochimal lines, the difference between a sea and continental climate would be included in the general view. These isothermal lines do not follow parallels of latitude. A study of the globe will reveal the details as to how the annual rotation and oblique motion of the earth in relation to the sun fix the *Alternatives of seasons.* tropical limits of the sun's apparent declination south and north of the equator, producing alternate winter and summer on either side of the line. The phenomenon well understood, it will be evident that the mean annual temperature obtained at different latitudes must decrease from the equator to the poles. Were the whole surface of the earth uniform, presenting a monotonous surface to the sun, unaffected by disturbed causes of mountain ranges, deep valleys, chains of lakes, expansive deserts, great forests, it would receive a uniform and equal degree of radiant heat. In that case the mean temperature of every point would be in proportion to the radius on the parallel of latitude. But the mean temperature of places in the same lines of latitude differs very materially. It will be observed that the lines follow strange and devious courses, though they run generally parallel over wide expanses of ocean or desert country.

Local variations.

Isothermals.

It will be seen by the variations of temperature that causes are in operation in different localities affecting the mean temperature other than distance from the equator. A glance at the isothermals exhibits the fact that though remoteness from the equator decreases the mean temperature, localities are independent of this cause.

One of the causes of these great variations in the same parallel of latitude is the physical formation of the earth. The continent of North America in its general contour resembles a spherical triangle of which one side stretches along the Pacific, one along the Arctic, and one along the Atlantic ocean. There are two axes of elevation ; one on the Rocky mountains, the other on the Appalachian chain. From Labrador to California is stretched a great water-shed which extends into four slopes and eight river basins. Three great ocean currents sweep along the shores, one southward in the Pacific, another eastward in the Arctic, and a third northward in the Atlantic. The Alleghany and

Physical contour of North America.

Rocky mountain ranges divide the face of the country
into the Atlantic plane and slope, which is washed by
the Atlantic ocean ; the valley of the Mississippi, lying
between the Alleghany and Rocky mountains, watered
by the Mississippi and its tributaries, and the Pacific
slope, extending from the Rocky mountains to the
shores of the Pacific. This great interior valley begins
with the tropics and terminates with the polar circle,
embracing in its area at least three-fourths of the entire continent. One of the great slopes in this valley
contains an elevation of 14,000 feet and serves to check
the winds from the Pacific and bestow on this vast region an insular climate.

Contour of
Europe.   All Northern Europe is a great plane extending far
into Germany, over a large portion of France, most of
England, all Ireland, Sweden, and a part of European
Russia.

In all this region the peaks of Wales, Scotland, the
Norwegian chain, and the Ural mountains, constitute
the only important elevation. Farther south the Pyre-
nees and the Alps exercise an important agency upon
the climate, lowering its temperature considerably. General causes tending to lower the mean annual temperature are the following : elevation above the level of the
sea when not forming part of an extended plane, the
vicinity of an eastern coast in high and temperate latitudes, the compact configuration of a continent having no
littoral curvatures or bays, the extension of land toward
the poles into the region of perpetual ice, without the
intervention of a sea remaining open during the winter ;
mountain chains whose mural form and direction impede
the action of warm winds. A comparison of the phys-

ical conformation of the coasts of Europe and the eastern part of the United States will show that the chief
agencies which control climates are directly the reverse *Reverse agencies.*
in each country.

Mention has been made of the fact that surplus heat
at the equator is impelled northward and southward.

Our planet is surrounded by a shell of atmospheric
density which acts as a bar to the escape of heat into
indefinite space. Cold air from the poles constantly
flows toward the equator to take the place of the ascending currents of hot air. The greater the difference
in degree of temperatures of Arctics and tropics, the
stronger these currents are. The Antarctic region has
for long ages been colder than the Arctic. The result *Polar currents.*
has been that currents from the south pole have been,
for so long a time, so strong as to move the surface
waters of the vast area of southern seas toward the
equator, producing the trade winds and ocean currents.

The tropical trade winds give both temperate zones *Trade winds.*
west and southwest winds, which are land winds to the
eastern coasts and sea winds to the western.

Consequently, the prevailing winds in Europe and
America are westerly. These land winds in the United
States tend to give a high summer, and a low winter
temperature, while the same wind, after crossing the
Atlantic ocean, bestows upon Europe a much milder
and more equable temperature.

To this cause may be attributed the mild climates of
Ireland, England, Jersey, Normandy, and Germany,
quite in contrast to the climate of interior Europe.

Thus, the myrtle blooms simultaneously in Ireland
and Portugal. In the Orkneys, in the same latitude as

Stockholm, the winter temperature is higher than at Paris. At the Faroe Islands the inland waters never freeze, owing to the moderating influences of the west sea winds.

Ocean currents. Ocean currents have an intimate and important connection with the climate of adjacent coasts. The gulf stream contributes essentially to the further modification of the climate of Great Britain, giving Ireland her perpetual green verdure, and England her perennial hedge.

A similar stream originating far south of Japan greatly modifies the climate of the Pacific coast.

Moisture exercises an important effect upon climate.

Taken as a whole, all the greater slopes on the American continent descend easterly toward the Atlantic, while the abrupt ones rise on its western aspect. This general configuration necessarily gives to Europe a moister as well as a more temperate climate than that of America in the same parallels of latitude. This would *Gulf stream.* be much more obvious but for the Gulf stream and its accompanying trade wind. From this source the Atlantic coast and the Mississippi valley derive a large portion of their moisture. The trade wind, fresh from the Gulf stream, not only extends its favor along the whole Atlantic coast, but reaches over to the Alleghanies, laden with vapor. Were the interior valley of the Mississippi exposed to the Pacific winds, it would undoubtedly have a much milder climate than it now possesses; and were the trade winds to cease their supply of vapor it would soon change from a fruitful region to a land so barren as to unfit it for the abode of man. For as the climate is, so is the country which it governs; and as climate and soil are, so is man. Upon these primal causes depend the character and extent of human civilization.

Use the Tellurian without the Day Circle, rectified for Dec. 21st, our northern winter.

The red and blue lines following such strange and Illustration. devious courses are the Isothermals and Isochermals; the red indicating average temperature of regions through which it passes during June and July. The blue indicate the average temperature for Dec. and Jan. The blue lines which traverse the Torrid zone indicate a high degree of heat averaging 80 during our winter, showing the effect of vertical rays. The blue line 70, south of Capricorn, 40 degrees from the equator, shows a high degree of temperature, it being their summer. But the blue lines 40, 30, nearer the Antarctic Circle, Compari-sons. denote much lower temperature during their summer than the red lines 40–30, at the Arctic Circle during our summer, and they are much farther from the south pole, clearly proving the greater frigidity of the southern pole. North of the equator the blue lines 10, 20, 30, 40, run nearly parallel across our southern States, diverging only when past mid-ocean and well toward the western shores of Europe. Line 40, as an example, gives the same winter climate to Weldon, N. C., Bristol, England, and Rome. At the same time south of the equator, it being summer there, Rio Janeiro, Port Natal, and the central belt of Australia have a mean temperature of 80. Thus, while blue lines show a high temperature south Greater ac-cumulation of the equator, they show a low degree north, illus- of cold at south pole. trating at a glance the difference between the effect of vertical and oblique rays, as modified by land and sea areas. The difference between an Antarctic and an Arctic summer is seen by glancing at the blue line near-est south pole, marked 30. Notice how far from the pole it runs, then observe how much nearer the north pole the *red* line 30 runs. The trade winds from the south pole are so strong that they impel a mass of water northward through the South Atlantic basin, up to the equator. A portion is deflected southward again by the jutting coast of South America, but the greater mass is impelled across the tropics northward, following the shores of the Gulf of Mexico; thence along our eastern coasts, finally reaching the very confines of the Arctic seas. The parallelism of our Gulf stream, as it flows northward, and the blue lines covering it, reveals the fact that a volume of heat ascends from a low land latitude to a high sea latitude.

Let us suppose a person whose health was such that he required a climate of 50 degrees Fahrenheit. He would leave Sacramento, crossing the continent to New Orleans and Mobile, embarking at Savannah for Gibraltar, taking a detour northward to avoid the heat of the Bermudas and Azores.

At Gibraltar, preferring a sea climate, of 40 degrees, our climate hunter must needs pack and fly by fast dispatch 1,500 miles to the Orkneys, where, let us hope, he reaped the benefits of the friendly Gulf Stream. In the spring he could go to Norway, thence to Spitzbergen for a short summer.

Observe the red lines which indicate the average temperature of the earth during July.

A high rate in the tropics — it is always hot there. While the mercury would rise to 80 at Cincinnati, Tripoli, Rhodes, and Pekin, a singular fact is noticed at California; the southwest trade winds give the coast a comfortable temperature of 60, but east of the Sierras

Marginal notes:

Wind and water.

Summer climates.

it takes a sudden rise of 10 degrees. A short distance farther east another rise of 10 degrees, inflicting a torrid atmosphere upon the vast continent lying between the eastern base of the Sierras and the Atlantic coast, from Montana diagonally across to Savannah. This elevation of the isothermal 80 illustrates in a marked manner the fact that land interiors are hotter and colder, alternately than sea levels in corresponding latitudes; and that western continental coasts have a milder climate than eastern coasts.

Have the climates of zones always been as they now are? No; the climates are constantly and slowly changing. Geologic changes.

What causes these changes?

The answer is found in the *Precession of the Equinoxes.*

It has been laid down that the north pole always points to the north star; the earth's axis always inclines in one direction; that the poles are always parallel. Practically. this is true; life is short, and man is principally concerned with to-day and to-morrow. Geology reaches backward and tells us what has occurred. Astronomy probes the past and future and prophecies for distant coming ages.

Astronomers teach that the axis of our earth projected to the heavens would now reach a point within $1\frac{1}{2}$ degrees of the north star; having approached it 12 degrees since the ancient astrologers drew their magic circles. Poles of the equinox.

Illustrate with the tellurian, rectified for March 20th. Rotate the globe, bringing the plane of the eliptic to a horizontal position. Let the solar index point at the prime meridian when the ecliptic crosses the equator.

Loosen thumb-screw C; gently lift the globe by grasping the arm B C, withdrawing it from its fixed position. Press the arm at B, slightly moving it eastward, or to the right, without rotating the globe on its axis. Observe that as you move it by gentle hitches the point of the *Precession of the equinoxes.* solar index slowly recedes. Each recession of one degree is equal to the change in polar direction in 71 years; thus, if the vernal equinox occurs March 21st this year, in 71 years it will occur March 20th.

The equinoxes have already receded 30 degrees since the earliest data, therefore the signs of the zodiac, which are arbitrary, are at this time a month behind their constellations. Turn the globe slowly as described entirely around, returning to correct position; this movement of the poles of the axis entirely round the poles of the ecliptic occurs once in about 26,000 years.

Illustration. Illustrate in another manner; rectify the tellurian for Dec. 21st; rotate the arm, describing the annual orbit of the earth around the sun, observing that the poles always point in the same direction. Lift the whole instrument and twist it sidewise, without disturbing its parts. The polar direction has been changed. Continue to displace and replace until you have completed an entire rotation of the circle of the zodiac. This illustrates the precession of the equinoxes through an entire circle once in 26,000 years; the earth meanwhile having made diurnal rotations to the sum of 365 times 26,000.

The orbit of the earth in its course through the year is not a true circle, but an ellipse; the position of the *Earth's orbit elliptical.* sun being slightly removed from its center. In June the earth is at its greatest distance from the sun — in *aphelion;* in January it is nearest the sun — in *perihelion.*

A line piercing the centers of both sun and earth, ex- Line of Apsides. tending through aphelion and perihelion, called the Line of Apsides, will, if continued indefinitely in both direc- tions. pass through the *signs* Capricornus and Cancer, but the constellations are not there; 2,000 years ago they were. for at that period signs agreed with their constella- tions. The slow change in direction of the Line of Apsides has caused the astronomical discrepancy. The line of apsides is also coincident to the plane of the ecliptic. As a result of precession we know that aphelion Relation to is slowly moving eastward, perihelion westward. Do our cli-mates. you ask, what have all these indefinite astronomical facts to do with climate on our own little planet ?

Everything.

Creation is a vast scheme of cause and effect, whose First cause. farthest nook and remotest corner is subject to the inex- orable mathematics of God the First Logician.

At the present time, the earth's perihelion occurs in Perihelion. January. As a consequence the south temperate zone is hotter at that season than is our north temperate during our summer. The speed of the earth in all its move- ments, when in perihelion. is greater in ratio of its greater proximity to the sun ; causing a shorter summer to the southern hemisphere than is our northern sum- mer six months later. In fact, though the summer of perihelion is hotter, its brevity prevents the reduction of south polar cold in the same degree as at the north pole during our longer season. Thus the accumulation Greater cold at south of cold at the Antarctics is not only proportionately pole. greater, annually, than at the Arctic, but it also in- creases in amount each year. The isothermal lines, as seen on the globe, show the relative accumulation of cold at the polar circles.

5

To recapitulate :

Second cause. 1. The earth has two motions — axial or diurnal, orbital or annual.

There is a time at which the earth in its orbit, having passed through perihelion in winter, arrives at a position in relation to the illumined hemisphere which causes **Aphelion.** equal day and night : — or *equinox.* There is another time when the earth having passed its summer at aphelion approaches a like position at the *other side* of the orbit, called *equinox* — respectively Vernal Equinox and Autumnal Equinox. These two positions gradually recede — that is to say, the day when a line passing from the center of the sun to the earth's center would touch the equator is always the same day of the calendar and is in the same month and sign of the zodiac, *for the sake of convenience.* But, upon this day, that central line varies in direction to the distant heavens. Spring must come in March, for it is the *equinox* which gives us the date and names, but the constellations seen in spring will, at a remote future period, be seen in September.

Third cause. 2. The precession of the equinoxes causes a change in direction (toward the distant stars), of the line of apsides. This will eventually cause — not a change in the *names* of seasons, but a reversal of the *climates* of seasons ; summer will at some distant period be cold, much colder than our present winters.

Fourth cause. 3. Having learned that the Antarctic region, being coldest, exercises a preponderating influence in the causes of trade winds and ocean currents ; thus establishing to a great extent our present climatic conditions : we know that a reversal of polar conditions must cause a reversal of effects.

Time will come when our northern winter will occur at aphelion. Thus a long season of oblique rays and greater distance from the sun will cause such accumula. tions of frost at the Arctic region as to make *it* the prime factor of our climate. Natural conditions will then be reversed — north polar currents will be strong. est, the Gulf Stream will be unknown, and trade winds will be indicated on marine charts with the barbed arrows pointing the other way. *Reversal of climates at some remote period.*

Such changes are too gradual to take into human account. A state of things like that described cannot culminate — it is too slow. Even a reversal of present climates cannot occur until long after our civilization has ripened, decayed, and fallen in the order of nature. A northern winter in aphelion occurred before the dawn of history. Many thousands of years ago such a period covered most of the north temperate zone with oceans of ice and snow, but that was prior to the elevation of our highest mountains. Those Arctic floods are known to us as glacial epochs. Their slow descent drove the earliest known human beings, the River Drift Men, far southward, and they in their turn were exterminated by the Cave Men. The northern winter approached per-. helion and the Cave Men followed their favorite climate northward, where their sole surviving heirs, the Esqui-maux. still luxuriate in perpetual snow and a diet of grease. Undoubtedly at some not remote period the north pole will be reached, as the limit of decreasing accumulation at the arctics has not yet been passed. The torrid zone slowly extends northward, our temper-ate zone as surely approaching the Arctic circle. This process is so slow that perhaps the best illustration of its *No immedi-ate danger.* *Glacial periods.* *First men.* *Their heirs.*

effects may be found in the history of modern civiliza-

Climate and intellect. tion. Intellectual energy exists only under proper climatic conditions. Constant high temperature prevents a marked rise above barbarism. A high rate of human achievement is only compatible with temperate zones. Recall to mind the gradual northward advance, during **Vast transi-** the last 3,000 years as recited in history. From Thebes **tions.** to Athens, from Athens to Rome, from Rome to Paris, London, Berlin, New York.

Civilizations rise and fall; races of men appear only to disappear; each epoch of human transition being allied to, if not coincidental with, a change in terrestrial climate.

Longitude and Time.

Longitude is the angular distance between two meridians, measured upon the equator or any parallel, usually **East and west longitude.** reckoned east and west from the prime meridian. Distance to the right is called east longitude; to the left, west longitude. Every point on a given meridian has the same longitude. Any two meridians include an arc having the same number of degrees whether measured upon the equator or a parallel. But the absolute length **Solar circles.** of the arc of a parallel is in proportion to the radius of its circle. Thus, in regard to the sun's declination at equinox, an observer at the equator would see a circle described from east to west through zenith; while the circle as seen at the arctic zone would be very small and far south of east, west, and zenith; though the time of the sun's course from rising to setting would be 12 hours at either place.

Distance, as expressed by the term "longitude," is measured on the earth in degrees, minutes, and seconds.

Distance, as expressed by the term "time," is reckoned by months, days, hours. and seconds.

A year is the limit of time as applied to longitudinal distance, because the geographical division of longitude corresponds to the mathematical division of a year. The earth, as you observe on the globe, is divided into 360 degrees, the space between two meridians compris- Degrees and hours. ing 15 degrees which is equal to one hour of mean solar time, each degree being equal to four minutes.

Bring the Calendar index to March 20th, and the meridian of Greenwich to the solar index at J. Observe that longitude is reckoned east and west from this meridian.   It is noon at Greenwich, and, as the earth rotates to the east, all longitudes east must have *passed* the solar index, therefore it is later in the day at all places whose longitude is east of the prime meridian.

It is afternoon at Constantinople, evening at Calcutta, Variation of time during midnight and the beginning of another day at longitude a day. 180.   Rotate the earth and bring 180 to the solar index. It is midnight at Greenwich, noon at 180, morning at Calcutta, and evening at Constantinople.   Can noon and midnight occur every 12 hours ?   Yes, midnight occurs every instant, and noon is perpetual.   Mark the night shade opposite O on the meridian of Greenwich at midnight.

Rotate the globe and observe that the point of noon under the solar index and midnight on the night shade are fixed while the entire map of the globe passes these points.

Time is fixed in space and variable upon the earth, for it is always noon under the vertical ray and midnight 180 degrees distant in either direction.   Bring the prime meridian to the solar index.

Hypothetic-
al journeys
of pigeons. *If the earth remained motionless* and a carrier pigeon were given wing at Greenwich for Chicago, he would have to fly westward 15 degrees an hour for 6 hours to reach Chicago at 6 o'clock, and a further journey of 6 hours at the same speed to reach meridian 180 at midnight.   Continuing his flight would he not see the sun rise in the west at 6 o'clock in the morning at Calcutta, having seen it sink in the east the previous night at Chicago ?

The bird flew in one direction the entire journey in 24 hours and saw the sun in opposite directions at nightfall and morning.

The carrier pigeon exactly describes the *apparent* motion of the sun.

He can make a much easier tour of the circle in 24 hours in fact, for he has only to perch at Greenwich and the earth's axial movement will carry him around without effort ; in this case the sun will rise in the east and set in the west.

Let us suppose a pigeon on wing at O, noon, March 20th.

The earth would make an entire revolution under him in 24 hours and he would perch hot, tired, and exhausted after a noon of 24 hours duration.   Imagine him taking flight the next day at noon from Greenwich for Chicago : Arriving there after a three hours' journey of 15 degrees an hour.   What time would it be in Chicago ?

Nine o'clock in the forenoon !

He arrived before he started !

When it is noon at Greenwich it is 6 A. M. at Chicago.

The pigeon flies westward *away* from the sun one-half the journey ; rotation of the earth moves Chicago *toward* the sun the other half, or three hours.

Resting at Chicago three hours, the bird sees noon for the second time after a duration of only 6 hours.

Starting again he flies eastward for home, arriving at noon the next day ; how far did he travel ?

It is but 90 degrees from Chicago to Greenwich, yet the pigeon flew 360 degrees ; for did he not fly the entire circuit of space from the solar index back to the same ?   At what rate did he travel ?   He could have made the journey easier by flying to the north pole in 3 hours, and perching when it was 3 P. M. at Chicago, and 9 P. M. at Greenwich.   Waiting until it was 9½ A. M. of the following day, at Greenwich, he would fly southward 2½ hours and perch at home after a 5½ hours' flight under perpetual noon and a rest of 18½ hours.   Or he might have flown from Chicago to the south pole, arriving there when it was 9 P. M. at Chicago, and 3 A. M. at Greenwich.   But his journey must have been continuous for he would have to start on his return instantly and gain a half hour to perch at Greenwich at noon.

Let two pigeons start from the equator at O, noon, March 20th, one going eastward the other westward at the rate of 15 degrees an hour, taking wing from a ship in the Gulf of Guinea.   They would meet at midnight — where ?

They would alight on their perch, whence they had flown.   Resting six hours, they start again, one flying from sunrise, eastward, toward noon ; the other flying westward toward midnight.   In 12 hours would they not meet on perch at sunset, March 21st.   Trace their journey with the globe, then map out the same journeys by men traveling on the earth at the same rate of speed, and observe the difference between the variation of time by traveling eastward or westward.

To find the Longitude of any Place.

Bring the meridian of the place to the solar index ; — the degree east or west of O is found where the meridian crosses the equator.

To find the Latitude of any Place.

Bring the given place to the solar index adjusted so as to cover it ; rotate the globe and find the degree indicated on meridian 170.

Sidereal and Solar Time.

A sidereal day is the interval between the moment a *fixed star* is on a meridian, and the moment when it is next on the same meridian. A solar day is the time from noon of one day to noon of the next day, or when the sun is on the same meridian. Illustrate with the tellurian rectified for March 20th — solar index at Prime Meridian *and in line with a distant window which stands for a star.*

Illustration.   Rotate the globe once, *exactly,* on its axis without moving the arm. This illustrates a solar day, and if the earth had only an axial and no orbital motion you will perceive that the solar and sidereal days would be alike, because your solar index would remain in line with your window. Now move the arm slightly and you will observe that the solar index does not point at the same meridian. Call it one day's interval, and explain that as the earth has made one exact rotation on its axis, the star only is on the meridian ; showing a sidereal day. Must not the globe be moved still more to bring the meridian under the solar index ?

During one diurnal motion of the earth on its axis, the *orbital* motion carries it eastward nearly 1° ; there-fore, it must perform 1° more than one exact rotation to bring the same meridian to noon of the second day — or between two successive upper transits of the sun. Or, to repeat in other words, if the sun and star cross the meridian together at noon to day, to-morrow the star will cross about four minutes before the sun. This variation extended through a whole year, makes it necessary for the earth to perform one entire revolution more on its axis to make solar days than would be required for the same number of sidereal days. This fact is made appar-ent by your use of the tellurian, for you observe that by each slight movement of the arm, you separate your star meridian from your sun's meridian ; and if you allow the globe to rest, it is necessary to make an entire circle of the arm before you can bring your star and sun again in conjunction.   .

Variation caused by orbital motion.

Conjunc-tion.

Second con-junction.

The solar day does not always differ from the sidereal by precisely the same amount because of the unequal rate at which the earth moves at different times of the year, in its orbit at the plane of the ecliptic. Time measured by a meridian sun is called *apparent* time. The average length of solar days for the year is *mean* time— the mean solar day ; thus constituting the *civil* day of 24 hours, beginning at midnight with the sun at lower transit. It is divided into two periods, each of twelve hours ; from the lower to the upper transit — midnight

Apparent time.

Mean time.

Civil day.

to noon — and from the upper to the lower — noon to midnight.

The interval by which *apparent* time differs from *mean* time is called the *equation of time.* The sun's change of right ascension is sometimes faster than if it moved on the equator, and sometimes slower ; therefore, the equation must sometimes be added to, and sometimes subtracted from, apparent time. Its greatest additive value is $14\frac{1}{2}$ minutes about February 11th and its greatest subtractive value $16\frac{1}{4}$ minutes about November 3d. The equation of time is zero ; mean time and true time are the same four times in the year ; Apr. 15th, June 15th, Sept. 1st, and Dec. 24th.

Local time is the time of day (solar or sidereal) on any given meridian at any instant. The beginning of a day is determined by the transit, across a given meridian, of the sun or star which has an apparent diurnal motion of 15° per hour ; therefore, the local time at two different meridians at the same instant must differ at the rate of 4 minutes for each degree, or 15° every hour. Since the diurnal motion of the earth is from west to east, it follows that time is later at places east of a given meridian and earlier west of the same meridian.

The Moon.

Our satellite resembles the earth in form ; in size it is one-quarter as large ; in volume. one-fiftieth. Seventy million moons could be stowed within the enormous bulk of the sun ; yet the apparent size of sun and moon is nearly the same. This is owing, of course to their relative distances, the moon being less than 240,000 miles away, while the sun's rays reach us across the abysmal depths of ninety million miles. You may reduce the apparent sizes of large and small objects by varying their distance from the eye. The dark spots on the moon are the yawning craters of extinct volcanoes ; the larger darkened areas are shadows of high mountains. The moon has been condemned as a defunct orb, without moisture. atmosphere, or life ; but late research claims to have discovered signs of vegetable life.

It is not known whether this condition is due to infancy or old age.

Civilization, with its telescope, calculus, and spectrum has not existed long enough to observe a great geologic change.

The sidereal revolution of the moon around the earth Lunar days. is once in 27⅓ days, but as the earth's orbital movement carries it constantly forward, the " face " of the moon is at the same point in relation to the sun once in 29½ days. This phenomenon is similar to the distinction between the sidereal and solar day as illustrated with the tellurian.

The moon always turns the same side toward the earth, therefore it turns on its axis once a month. One-half its surface is illumined 15 days by the sun and becomes, by accumulation of heat, much hotter than our torrid zone.

At the same time the dark half has ample time to radi-ate its heat and accumulate a correspondingly intense degree of frigidity.

CHAS. W. HOLBROOK'S LUNARIAN.

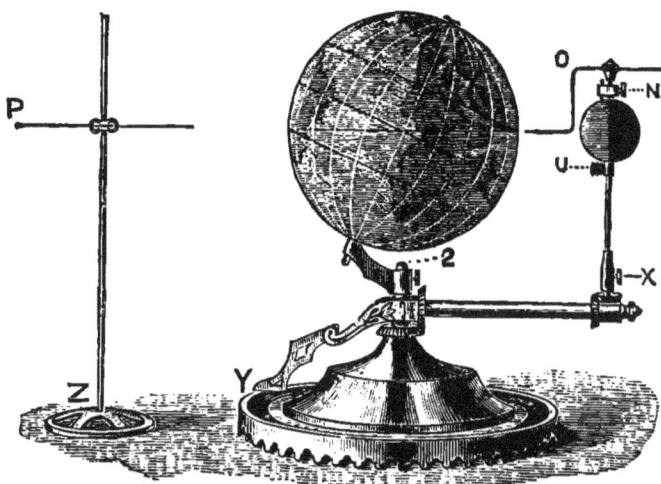

Cut No. 2.

Description. The Lunarian is adjusted for use as shown in the cut. Mount the globe at the center of the base. Screw on the moon *precisely* as follows: Loosen thumb-screw X, *tighten* thumb-screw N, remove the pointer, or lunar index O; press the collar down over the post at X, take hold of the cross-piece at N and turn to the right, pressing downward. When you have screwed on the moon tighten thumb-screw at X. Rectify for Dec. 20th by bringing the calendar index Y to the winter solstice. Place the sun Z opposite Y, let the solar index P point at meridian 95.

If you will now mark one side of the moon, and turn that side toward the sun, calling it the "face" of the moon, screw up tightly at N, adjust the dark shade away from the sun and screw up at U, the apparatus is ready for use. To show the moon receiving the light of the sun above or below the earth, hold it at a distance from the earth, in an opposite direction to the pupils. Rotate the arm toward the west and observe: —

1. The "face" of the moon, whether visible or invisible always points toward the earth. *Illustration.*

2. The "face," to be visible, must be directed toward the sun. Some portion of the "face" half of the moon is always visible from some place on the earth except at eclipse. The sun forever shines upon and irradiates a half of the moon except when the moon is eclipsed at "full," on the dark side of the earth.

3. The face of the moon disappears by *rotating into* darkness, and though constantly facing the earth is, for a portion of time during each lunar day, invisible by its own shadow. The black shield on this mechanical moon well illustrates the obtrusion of a lunar night by the intrusion of the moon's surface into darkness. *A perfect mechanical moon.*

4. The amount of luminosity visible to an observer depends upon whether the moon's course is during day or night. If during the day the sun's light extinguishes the lunar orb — if during night the sun's light is reflected from the lunar orb. The observer will, in the latter case, see the moon; the visible area being in ratio to the parallelism of his line of vision to the solar rays which illumine the moon in opposition.

Astronomers tell us that the moon's orbit is inclined to the plane of the ecliptic 5 degrees; which means that while the earth goes swimming along the plane, the moon, like a flying fish alternately leaps above and dives below the level. Would not the effects be the same if the moon rode horizontally and the plane vibrated with a wavy motion once a month ?   So far as our mechanical illustrations of the moon's movements are concerned, the

Increased obliquity of the plane of the ecliptic.

effect is the same.   Remove the earth and incline the plane of the ecliptic by inserting projection 2 (cut No. 3) at the aperture 2 (cut No. 2.)

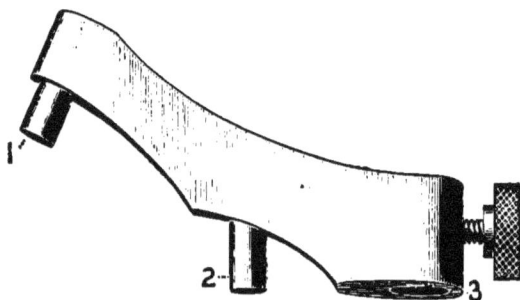

CUT No. 3.

Rotate the arm, and observe that, if you have the earth inclined with the north pole turned farthest from the sun, the moon is in perigee, nearest the earth, during the night; in apogee during the day, in the winter season.   Insert the lunar index O, and it will trace upon the earth a line running 5 degrees north of the plane of the ecliptic and the same distance south, illustrating, *exactly*, the fact that the maximum variation of moon's point of sunrise and sunset during the year is 57 degrees, its minimum variation being 37 degrees. This fact is apparent when we reflect that owing to the inclination of the earth's axis, $23\frac{1}{2}$ degrees, the variation of sun's place on the horizon is 47 degrees; and the obliquity of lunar orbit 5 degrees above and below may add or subtract 10 degrees to and from 47 degrees.

The Phases of the Moon.

Have the globe properly adjusted, meridian 95 directly above Dec. 21st.

Place the sun Z at March 20th, and adjust the solar index P to indicate the Prime Meridian at the equator. Adjust the moon shade and bring the Calendar index to November which is your point of observation. Look across the curve of the globe and by rotating the arm you can describe all the phases of the moon.

High and Low Moon.

Rectify for December by placing the sun in its first position. Rotate the globe and bring the United States to the dark side of the earth.

Rotate the arm, calendar index at December. Adjust the moon shade to show "full" moon at night. Rotate the arm and observe that the lunar-index indicates latitude 28 north, during the night when the moon is visible, descending to 28 south during the day, when the dark side is toward the earth. This is a "high" or winter moon. To show the "low" summer moon, rotate the earth on its pivot at 2, reversing its position; or let the earth remain, and place the sun at June, the moon at December with the dark shade reversed. In this position the moon is only visible when low. Variation of plane of lunar orbit is very slight. The difference between high and low moon is caused by the change from winter to summer, of the position of the northern hemisphere in relation to day and night.

Perigee and Apogee.

The lunar orbit is elliptical in form. Lunar apsides, an imaginary line extending from the moon's place when at greatest distance from earth to the nearest limit in perigee, has a very eccentric motion. This line is never straight and seldom touches the earth. It makes a complete revolution once in nine years. Apogee and Perigee are in opposite signs of the zodiac but do not bisect the circle. The moon's apogee in the constellation Aquarius in January, 1878, made the circuit of the zodiac by January, 1887, its motion being quite uniformly vibratory; but perigee darted forward and backward with strides which covered one-sixth of the circle or more.

The effect of perigee is seen in a total eclipse of the sun — apogee causing an annular eclipse. The lunar apse revolves in one-half the time required for the saros.

Harvest Moon.

As the moon performs her journey around the earth in a month, the daily advance eastward is 12 degrees. If the lunar orbit were in the plane of the ecliptic it is plain that we should have two eclipses during the month, but we see our mechanical moon tracing an orbit which is part of the time above, the other part below the ecliptic.

Difference of angles between the plane of the lunar orbit and the equator at Aries and Libra. If you will rotate the arm and move the moon slowly, you will observe that at Capricornus and Cancer the lunar orbit is nearly parallel to the plane, and equator. When following this course the moon's rising for a few days is at nearly the same hour.

But when the moon approaches Aries and Libra its

path is much inclined to the equator; at Aries ascending to the equator, at Libra, descending toward the equator. When the moon *ascends,* three or four successive risings will occur at nearly the same hour, because the horizon of an observer in latitude 45 will be carried toward it at a slight angle.

When descending in Libra, the moon will rise at nearly the same place but at long intervals, for a few nights, because the horizon approaches moonrise more obliquely.

To illustrate: Bring the moon to Aries and rotate the earth on its axes, observing that your horizon approaches the moon's path as it *ascends* to the equator. Whereas, when the moon is full at Libra, at vernal equinox, your horizon and the moon's path diverge. In the former case moonrise occurs at nearly the same hour at *places* widely separated; in the latter case, moonrise occurs at nearly the same point on the horizon but at *hours* widely separated.

When the moon is full in Aries the sun is in Libra, the autumnal equinox is at the eastern horizon. This is the harvest season in England where the name was given to the harvest moon.

Nodes.

Adjust the lunarian as described; the globe, as explained by Cut No. 3. Plane of the ecliptic is inclined in such a manner as to dip its northern and elevate its southern paraliel. Rectify the lunar index to cross the ecliptic at the prime meridian, as you rotate the arm. The moon's orbit crosses the ecliptic. It will cross again at meridian 180.

6

These points are nodes.

Each sidereal revolution of the moon comprises two nodes. The *ascending node* is a point on the ecliptic where the moon crosses in passing from below. The *descending node* is the point crossing in descending. Bear in mind that *nodes do not relate to the equator*, but to the plane of the ecliptic. As the lunarian is arranged, the ascending node occurs March 20th, the descending, Sept. 23d.

If sidereal and synodic revolutions were identical, nodes would always occur at the same points. But the moon makes two nodes in $27\frac{1}{3}$ days, the orbital progression of the earth impels the satellite forward, changing position so much that the moon must advance $2\frac{1}{4}$ days farther in orbit to recover the same relation to the sun, or the same *phase*.

From ascending node to ascending node is a *nodical revolution* of the moon.

From full moon to full moon is a *synodical revolution* of the moon.

Nodes relate to the moon's place as seen from a star, and they occur only at the plane of the ecliptic.

Phases relate to the appearance of the moon as seen from the earth. Phases occur every $29\frac{1}{2}$ days, while nodes, occurring oftener, must vary in position. They do, in fact, recede about 19 degrees during one year. A given node passes around the entire circuit of the ecliptic in 18 years. The place of nodes are sometimes between the places of perigee and apogee; at other times are identical to them.

Illustrate with the Tellurian: Rectify for December as shown in cut No. 1.

Incline the upper half of the day circle by bending it away from K toward and beyond L, letting it rest 5 degrees above the plane of the ecliptic, after bringing the latter to a horizontal level. The hoop represents the half orbit of the moon above the ecliptic. If the hoop were extended on the same plane entirely around the globe, it would represent the entire orbit of the moon. The pivots or nodes of the hoop describe the two points when the orbit crosses the ecliptic. Rotate the arm slowly, observing that

1. If the moon's orbit cross the ecliptic twice in $27\frac{1}{3}$ days, it would make during the year thirteen each ascending and descending nodes, and a fraction over. This fraction gives the moon in the year 1888 fourteen ascending and thirteen descending nodes.

2. As the *phases* of the moon are governed by its position to the sun, it follows that the nodes and phases are irrelevant to each other.

3. That a node is not necessarily made at the equinoxes, or at any other given point.

4. That a "high" moon may be 5 degrees below the ecliptic, a "low" moon 5 degrees above the ecliptic; the terms "high" and "low" relating to the equator only.

5. That a high moon may traverse a low node, that is, a high moon is not necessarily at the highest point from the ecliptic, and *vice versa.*

6. That the moon passes through the signs of the zodiac once each sidereal revolution.

7. By the Lunarian. Cut No. 2.

That the sun is in the same sign as the moon when the moon is new; that the earth as seen from sun or moon would always appear in the opposite sign ; that the earth and sun are in the same sign when the full moon is eclipsed, but not when the moon is full. The full moon is never exactly in line with the earth and sun. When the sun sets in the west the moon may be in the opposite sign, but not at an opposite point — if the sun set Sept. 23d at the first point of Libra, the full moon will rise in the last of Aquarius.

The straight line from setting sun to rising full moon does not pass through the center of the earth nor across its edge at the feet of the observer, but at a distance from the earth.

Eclipses.

1. When the moon is between the earth and sun it is "new" and invisible.

Total and annular eclipses.
Illustrate by placing the moon in this position, the shade hiding the moon from the earth. If the new moon approach a node the sun will be eclipsed. If the moon is farthest from the earth, in apogee, the eclipse is annular. If the moon is nearest the earth, in perigee, the eclipse is total. A dime held before your eye will eclipse wholly or partially a large, distant object if you vary its distance.

A dime placed upon a nickel will illustrate an annular eclipse of the nickel. A dime placed upon another dime will totally eclipse it. The moon is so far away in apogee, that it will not entirely obscure the sun's disk, and the latter will display around the moon an annular ring of light.

The shadow of the moon is conical in form, the apex of which will barely reach the earth, while, with the moon in perigee, this shadow will more than reach the earth, forming a dark spot about 180 miles in diameter.

The sun is so large that if the moon approach within 15 degrees of a node there will be a total eclipse, visible to some portion of the earth ; if it approach within 30 degrees the eclipse will be partial. The total shadow called "umbra," of the moon, is small at the earth's surface, compared to the partial or "penumbra" shadow, which is about 4,000 miles, so that an observer 2,000 miles from the scene of a total or annular eclipse may witness a partial eclipse. It is possible for an eclipse Both at the same time to be both annular and total to different observers, for possible. example, the eclipse of March 5, 1886, visible as annular at Tampico at sunset and New Guinea at morning, might be total to an observer at Kingsmills, meridian 180 — at noon, because the latter place is 4,000 miles nearer the moon than the former, and the moon would appear $\frac{1}{80}$ larger. This phenomenon could only happen when the relative distances of moon and sun at syzygies were such that the moon's shadow would be *too long* to reach the *nearest* part of the earth, and *not long enough* to reach the earth's center. The moon's shadow travels across the earth at an absolute velocity Velocity of moon's of 2,000 miles an hour ; sufficient to carry it across the shadow. earth's disk in four hours. The rotation of the earth reduces this one-half at the equator where the shadow falls vertically. Over the curvature of the earth at morning and evening the shadow passes more quickly; also at increased latitudes, on account of slower motion.

To our human vision the sun and moon differ so little in size, and the velocity of the shadow is so great that the duration of total and annular eclipses is short, lasting but a few minutes at one point of observation.

A solar eclipse begins on the western edge of the disk, the moon approaching the sun always from the west, in the same manner as a meridian of the earth approaches, *overtaking* the sun, and passing across its disk to the east. A spectator at the moon, during a solar eclipse, would observe a small black spot moving eastward upon the earth, this spot surrounded by a much larger round shadow.

Umbra and Penumbra.

The spot is the umbra, the shadow the penumbra ; the umbra is the total obscuration, the penumbra the partial obscuration. The sun is not affected when eclipsed — we are simply unable to see it.

Umbra and *penumbra* of the moon during a solar eclipse may be illustrated with the Lunarian as follows:

Place the sun a few feet from the earth, the moon between them. With the flat of a knife-blade or paper-cutter, press a cord against the earth opposite the lunar index. Draw the ends of the cord backward away from the earth across the upper and lower edges of the moon, fastening them to the sun beyond. The cone between the moon and earth is the *umbra*, and the flat blade will describe the area of totality upon the earth.

Now draw the sun backward a little and hold the cord against the globe with the *edge* of the blade, bringing the cone to a point, describing the umbra of an annular eclipse. The penumbra will be shown by passing the cord from the center of the sun across the moon to the earth.

A good illustration of solar eclipse may be made with the Tellurian.

Illustration

Place the globe on the arm, as in Cut 1, put on the day circle and night shade. Take a position, on the dark side of the globe, at such distance that, with the moon in your hand, you may obscure the whole globe — a total eclipse. Extend the moon farther from the eye and the day circle will represent the annular ring of light which is seen during an annular eclipse. In this case the globe is the sun and yourself the earth, the umbra reaching your eye.

The area upon the earth covered by the penumbral shadow of the moon during an eclipse of the sun is called the *solar ecliptic* limit. It is so large that, if a lunar eclipse occurs very near a node, there must be one and may be two solar eclipses at preceding and follow-ing conjunctions. Thus, there may be as many as six eclipses while the sun passes the two nodes. Solar ecliptic limit.

As a result of the backward motion of the nodes another may occur during the year, the greatest num-ber of possible eclipses in a year being *seven*, of which five are solar and two lunar.

If an eclipse of the sun occur in passing each node, the lunar ecliptic limit is so small that the moon may escape an eclipse when in opposition, the previous and subsequent orbital courses carrying it above and below the shadow of the earth. Lunar eclip-tic limit.

Eclipses of the Moon.

In previous experiments with the Tellurian, concern-ing the phenomena of light and darkness, it was seen that the sun eternally illumines space in all directions. Any planet, satellite, orb of any degree, flying asteroid, or floating speck will, if opaque, have its one portion illumined, while its other part will be dark. Any orb receiving light from the sun will cast a shadow away from the sun. The earth moving along the places of the ecliptic casts a shadow which, like a comet's tail, always points away from the sun. This shadow moves at the same rate of speed as the earth, measured by degrees, its velocity being greatest at its farthest extreme does not affect the fact of degrees as distinguished from miles.

Illustration. This shadow of the earth may be shown with the lunarian, adjusted as in cut 2. Draw a cord around the globe, crossing the north and south pole, extending it beyond the moon four times the distance of the moon from the earth, bringing the ends together. This cone is the umbra, and the moon is eclipsed in the center of its diameter. Move the sun away four feet, and fasten a cord at the solar index, drawing the cord across the upper and lower edges of the earth and beyond. This illustrates the penumbra. Within the penumbra there is light from a portion of the sun only; within the umbra no solar light reaches.

Sizes of Umbra and Penumbra at the moon. The diameter of penumbra where the moon enters is about five times the moon's diameter. The diameter of umbra where the moon enters is about two and two-third times the moon's size. The edge of the umbral shadow is so indistinct — gradually fading into penumbra — that the moon, to be totally eclipsed, must pass its own diameter into the total shadow once and a half.

The area of totality for lunar eclipses is, therefore, very small. If the moon's orbit were identical to the plane of the ecliptic, the moon must pass through the umbra every month. There would then be a total eclipse of the moon when in opposition, and of the sun, when in conjunction. We have illustrated the orbit of the moon as inclined to the plane, and, at the moon's real distance (120 times the moon's diameter from the earth), it will readily be seen that opposition and conjunction are generally passed above or below the earth's shadow.

The earth and shadow move in the plane about one degree daily, but the moon advances about thirteen degrees daily. If the earth had only an orbital motion the moon would appear to pass quickly through the shadow. Owing to the axial rotation of the earth the progress of the moon through the shadow may be much retarded by the fact that an observer at the earth would, by his rapid eastward movement, keep the moon and shadow in occultation, thus prolonging an eclipse, from first to last contact, to several hours. *Velocity of earth's shadow.*

An eclipse of the moon begins on its eastern edge and ends on the western. When the moon overtakes the sun the sun is eclipsed, beginning on the western side; when the moon overtakes a shadow the moon is eclipsed from the eastern.

During a lunar eclipse the moon is entirely enshrouded, the side nearest the sun being eclipsed, and the farther side black with intense darkness. At such a time, a man on the moon would say that the big earth passed between him and the sun from east to west, and to him the earth would be an immense black spot. The whole firmament would glow with a subdued light.

Lunar Journey.

A pleasing and entertaining use may be made of the lunarian by following the phases of the moon as described in an almanac. (The writer, without wishing to be invidious, begs to say, parenthetically, that he found Ayers' Almanac to be the most exact and minute in its lunar records. It is quite on a par with the Nautical Almanac published by the U. S. Government, so far as it goes, and can be had gratis of any druggist or storekeeper.)

In following a lunar course, always place the sun opposite the date, for it is the sun's position in the heavens which gives us the constellation, month and date. You observe in the spaces divided into constellations three characters; a small astronomical *sign* at the left corner, the *name* of the constellation and a figure illustrating the constellation. We will ignore the signs, advancing by *names* and *subjects.* *

The *name* is position for *first* of space and the *subject* for *last* of space.

Adjust the lunarian as shown in cut No. 2.

First night. Suppose we begin with July 9, 1888. Place the sun in position opposite that day, the solar index adjusted to the plane of the ecliptic when horizontally level. New moon on last of Gemini. Rotate the arm to position directly above the twins, as they are described airing themselves in the balmy atmosphere of summer. Sun

Eclipse of the sun illustrated. and moon are in conjunction, rising and setting at nearly the same hour. The moon is, of course, invisible at the earth, as it approaches the solar ecliptic limit at ascending node. The approach to the node is from below the plane of the ecliptic, and, as the moon enters the limit of totality, its upper limb obscures the lower limb of the sun for a brief space of time, sending the shadow darting

Where visible. across the lower limb of the earth. In this case, the eclipse is visible only in the southern Indian Ocean, and is so brief that before the diurnal rotation of the earth can bring a southern continent into point of contact, the eclipse is over. This eclipse occurs when it is our night and when we are brought into daylight, the moon is extinguished by the greater force of the sun's light. We never see the moon except when it is in position to reflect to us the sun's light. If the moon be between us Invisible and the sun we might be able to see the dark side, as we moon. see the black disk of our imitation moon, but for the fact that the atmosphere in its prime capacity so thoroughly mixes and equalizes the rays of light as to prevent the sight of a new moon. So that, at an eclipse of the sun, the moon, as yet invisible, intrudes her black disk across the sun ; without sound or warning — appearing suddenly from nowhere and disappearing into the same nebulous limbo.

A new moon is invisible to all parts of the earth except at times of solar eclipse ; at such times it is visi- Visible new ble only at such places as are covered by the lunar moon. shadow. A full moon is visible to all parts of the earth, except when eclipsed, at which time it is visible but not bright — the degree of intensity of its obscuration depending upon whether the shadow be umbral or penumbral.

July 10th, second night.

Moon sets 70 minutes later than the sun, having Second advanced 12 degrees in its orbit (that difference in time night. constituting the difference between the moon's solar and sidereal days) in the first of Cancer. Place the arm over the *name*. On this night the slender crescent moon is Crescent visible a short time after sunset. Take a position where moon. you can look across the edge of the globe at the sun and you will discern the slender crescent-shaped rim of the moon, if you have properly adjusted the moon shade. You will notice that the moon appears to be nearer conjunction than on the previous night, when invisible. And

Nodes. so it will appear for another night, but you will bear in mind that the moon has crossed the plane, is now above it, and that the plane of the moon's orbit is not coincidental to the plane of the ecliptic (the earth's orbit) but inclined to it : also that this 2 inch moon at a distance of 20 feet and the sun a mile and a half away, would cast a shadow

Moon's shadow. above the earth. When the moon is near a node, the plane of its orbit is inclined to the plane of the ecliptic, and at the rate of speed with which the shadow advances, you can readily see how at noon of one day that shadow might touch the southern extreme of the earth, losing itself in space in an hour, because the eastward advance of the moon is so much more swift than the upward.

When the moon's course lies between two nodes, the orbital direction is not so oblique to the ecliptic but more nearly parallel, at which time the shadow moves in a more horizontal direction, as explained in observing the "harvest moon."

Third night. Third night — crescent a little wider and moon set still later. It has been said on a previous page that a high moon may run below the ecliptic and a low moon above, and that the terms "high" and "low" relate to the

High and low. equator only. You will find in your almanac that the moon runs high and low every two weeks alternately. Possibly the pupil may have gained the impression that the moon runs high all winter and low all summer, in which case there would be a season of medium declination, as in the case of the sun at the equinoxes. This is an opportune moment to illustrate these facts of different meridian altitudes of the moon's courses.

The apparent course of the sun is at the plane of the
ecliptic, yet the sun is high during summer and low
during winter, why then should not the moon follow
nearly the same course, since its path is so nearly the
same as that of the sun? Because the entire variation
of the sun's declination is accomplished during the passage of the earth through the entire circle of its orbit,
and this maximum difference requires a year, as shown
with the tellurian: whereas, the moon passes entirely
around the earth in one month. Illustrate with the
lunarian; sun and moon in conjunction July 10th. The High and
solar and lunar indexes both indicate a high latitude low moon.
upon the globe, and, in fact, both are high; but the
moon is not visible and in popular parlance "there is no
moon" Now advance the moon to July 16th — arm
above the subject of Virgo. Rotate the earth slowly and
observe that when the United States is at noon the moon
is one quarter circle beyond. As you rotate the globe
toward the east you will notice that the lunar index
which, only a week previously, gave a height of 25 degrees north now gives nothing. That is to say, on the
day of July 9th, the moon had an altitude equal to the
sun's, whereas, a week later her declination has receded
from the Tropic of Cancer to the equator; a variation
equal to that of the sun in three months' time. Advance
the moon to Capricornus — a difference in one more
week of 30 degrees; showing that the moon's altitude
is quoted for *night* season only, by popular voice, while
the astronomical signs are given for the two points of
opposition and conjunction. Thus, by pursuing this
branch of lunar study you will find that when the sun is
high the moon is low and *vice versa.* July 16th, seventh

night. Moon sets at quarter past twelve in the morning in the last of Virgo — sun at date — arm directly above the virgin. Adjust the day circle, observing that it divides the bright half of the moon in two parts, one

Quarter moon. part being visible to the earth. On that night the quarter moon is visible at sunset; at a meridian altitude, setting far to the southward. For several nights it rises later and "waxes" from the "quarter" to "gibbous" (which means more than half and less than whole), from gibbous to full, which phase brings us to July 23d.

Fourteenth night. Fourteenth night. Full moon at a node, in first Sagittarius — advance arm to *name* and place sun at date.

Full moon. As the moon approaches the descending node it enters the penumbral shadow at the lunar ecliptic limit, at 10 P. M.; reaching umbra an hour later.

We have learned why totality does not occur until the moon has passed more than its full diameter into the shadow, and in this case totality does not occur until 12 o'clock; lasting an hour and three quarters.

At half past two in the morning the moon passes out of the total into the partial shadow, and into sunlight an

Eclipse of the moon. hour later. The magnitude of the eclipse is 1.825 of moon's diameter.

Where is this eclipse visible ?

To the larger part of the world except Eastern Europe and Asia.

Illustrate with the lunarian; rotate the globe, bringing the United States into sunset position over Libra; sun at July 23d. Set the polar index to 7.30 at the circle of

Where visible. figures — not the numerals. Observe that at this hour of our sunset, it is near the time of sunrise at Eastern Europe and Asia, and at those regions it is past the time of moonset. It is midnight at Western Europe, and those countries are, by the rotation of the earth, within its shadow. The moon being also within that shadow they are enabled to see the eclipse. And we who reside in America are not able to see the eclipse until we too are rotated into the shadow. Before Eastern Europe and Asia can make the diurnal circuit and reach the shadow, or *night time*, the moon has passed beyond and the eclipse is over. Duration of the eclipse.

We have learned that all eclipses depend upon the position of the moon in its orbit; that they can only occur when that orbit crosses the plane of the ecliptic — viz.: at or near a node; that the nodes or places of crossing recede, on account of the advance of the earth along the plane; that eclipses occur only at opposition and con-junction of moon and sun.

It is plain, therefore, that if at the time of the eclipse just described the moon's descending node had occurred at Virgo or Libra, its course when reaching Capricornus would have been so far below the plane as to carry it below the shadow of the earth. In that case we would see the full moon rise far to the south, but no eclipse.

If the node at Libra were *ascending*, then the lunar course would be above the earth's shadow and outside the lunar ecliptic limit. Moon above or below the earth.

To resume the lunar journey; — after the full moon the illuminated disk appears to shrink, rising later each night until July 30th.

Twenty-first night — last quarter moon in first of Aries — sun at date. Place the moon over the *name* and note that at the time of our sunset the people of Arabia see the quarter moon nearly in their zenith while to those who Last quarter.

live in Prussia the moon is "low"; both countries see it until sunrise, whereas we do not see it until midnight. As the moon wanes from last quarter to new it is to us a "morning moon" and is visible until Aug. 4th, after which time the sun's light is too strong and it disappears. Now if the moon had made a node at quadrature, we would not see it again until Aug. 10th, but it is approaching a node and a possible eclipse.

Eclipse of the sun. Aug. 7th — New moon in the middle of Cancer — sun at date — place the arm over the crab and observe that, as the ascending node occurred Aug. 6th, the lunar course carries the moon so far above the plane that the lower limb crosses the upper limb of the sun, the shadow touching the earth at its north polar extreme ; causing a partial eclipse of the sun visible only to Northern Norway and Sweden.

Moon's shadow falls high or low. At the time of the previous eclipse, July 9th, the node occurred after the transit and the moon's course was far enough below the plane to send the shadow across the Antarctic regions. At the time of the eclipse of the moon July 23d, the node occurred in the middle of the shadow, at exact opposition, therefore, the eclipse was visible to nearly the whole earth.

Tides.

Solar and lunar forces. The earth is mostly covered with water. Both sun and moon exert great powers of attraction upon the earth. We have observed, in studying the causes of the precession of the equinoxes, how they pull at the equatorial protuberance, causing the earth to revolve. Their power of attraction upon water varies on account of disparity of distance ; the moon exerting the most force.

The effect of attraction is to raise the waters of the
earth above their normal level ; thus, if the earth were
entirely covered with water there would always be under
the vertical rays of the sun a wave, or swollen mass.
Under the moon the same thing would occur in greater
degree, and these waves would behave in a strange manner. The solar wave would remain still, the lunar wave
following the moon around the earth. At times of solar
eclipse, these forces would unite into one great wave.
When sun and moon were in opposition, or a lunar
eclipse should occur, there would be one great swelling
liquid wave on both sides of the earth, the lunar wave
being much the greatest. But neither sun, moon, nor
both combined could draw the water from the side of
the earth farthest from them ; there would, therefore,
always be *apparently* two waves. If sun and moon were
pulling together, their wave would be the greatest. At
the extreme limits of attraction the water would be less
deep, because it had been drawn away to make the wave,
while upon the hemisphere in darkness the water would
stand at a normal level. This is precisely what happens
in nature. The waves are called *tides*.

Illustrate with the tellurian ; place the globe upon the Illustration.
geared arm as in cut 1, and the moon upon the central
post at T ; put on the day circle with the hoop C bent
downward to horizontal beyond L. Place the night
shade on the side toward the moon, the sun at a distance
from the earth and on opposite side to the moon. The
moon will exert an attractive influence greater than the
sun, as the protuberance of the night shade is greater
than that of the hoop. Rotate the arm and the earth
will be revolved from west to east, the tides from east

7

Full and ebb tides.

to west. Wander along the ocean beach and you will mark the rising and falling tides twice a day. When at their height, they are "full tides," when at their low mark. they are "ebb tides." This refers simply to their daily action. When the moon is new, or full, the tide is higher than the daily average, and is called "spring tide." When the moon is in quadrature the water ebbs lower, and rises not so high as the average flow, and is called "neap tide."

The general action of tides is much modified by the configuration of the coasts which the approach, or rather *which approach them*. In mid-ocean, tides are not discernible. In gulfs having broad openings in the direction of the advancing tidal wave, tides will rise higher at the western shores, but will not ebb so much, leaving the relative distance between a full and ebb tide about the same. Gulfs having their openings at the west will not feel the effect of the tides unless they are large bodies, like the Mediterranean Sea, having sufficient

Variation.

volume to respond to the attractive force. In the sea just named, the tide varies but three feet; while the German Ocean receives the onflow of the Atlantic tides through two gates — the North Sea and English Channel — doubling the system and giving them four tides a day. The great variation of tide at Bay of Fundy has two causes. 1. The advancing wave across the Atlantic is increased by the opposing force of the Gulf Stream. 2. The increased wave, reaching the large gulf, is suddenly brought to head by the shore resisting and penning it in. The water cannot flow onward nor escape , it *must* rise. And it does, to such a height that the observer may see the tops of a ship's masts at wharf in

New Brunswick at low tide, and the hull of the ship moored to the same level a few hours later. Inland seas have little or no tide.

Tides are greater when the moon is in perigee than when in apogee. The earth's perihelion or aphelion also affects the tides. The greatest attainable height of a tidal wave must, therefore, occur under these conditions — viz., the earth at perihelion ; the sun totally eclipsed, at a node very near an equinox, by the moon in perigee.

Figures denoting tidal rise and fall are deceptive. For example, the rise of seventy feet in the Bay of Fundy is only a rise of thirty-five : the other thirty-five is the fall. Let 1 = the normal height of the water ; a wave of 35 feet in height approaches, followed by a fall of 35 feet to level and 35 feet more to make a rise at some other locality ; $35 + 35 = 70$. A vessel at sea will mount the crest of a wave thirty feet high, but part of the wave's measurement must be given to the trough in which the ship rides before mounting The trough is below, the crest above the normal level ; the measurement of both equals the amount of displacement, so to speak. Tides follow the vertical direction of the moon after an interval. Water, though elastic, does not respond to gravitation instantly, and in all ports of commerce the times of tides are tabulated and published.

Tides are considerably affected by winds. A tidal wave accompanied by a strong, steady breeze will be lower in mid-ocean and higher at the western coasts, for the wind will not only accelerate it but will *pile it up* and retard its ebb not only as to time, but as to quantity. A tidal wave opposed by a strong wind will be higher in mid-ocean and lower at shore ; its ebb will be accelerated and increased.

A wind blowing up a considerable river will cause a *rise* but not a tide, for tides are regular by cause and in effect. The writer, when a lad, was aboard a steamer which ran aground in the Potomac River during a neap tide accompanied by a strong breeze. The propulsive power of the wind raised the water to the height of a "spring" tide. The question was, would the next spring tide float the vessel ? It did not. The succeeding spring tide did, however, as it was accompanied by a strong breeze.

Tides and their changes are very important to commerce. As a rule, great bays are divided from the ocean by lines of shoals extending across their mouths. At normal tide a vessel may not be able to enter on account of her great draught. By awaiting a high tide she crosses the bar and rides safely in harbor where the water has greater depth. Ship masters love to sail with tide and wind, for then the sea is smoother and speed greater. There is nothing more uncomfortable than to round a cape against the wind and with the tide, for the sea is rough and dangerous with peculiar waves called "choppy." Against a rocky coast the incoming tide, if assisted by the wind, dashes and beats with great violence, sending the spray to great heights and thundering its tireless monologue.

www.ingramcontent.com/pod-product-compliance
Lightning Source LLC
Chambersburg PA
CBHW021945190326
41519CB00009B/1150